Transmission Electron Microscopy

IV Spectrometry

Library of Congress Cataloging in Publication Data

Williams, David B. (David Bernard), 1949–
    Transmission electron microscopy: a textbook for materials science / David B. Williams
and C. Barry Carter.
      p.   cm.
    Includes bibliographical references and index.
    ISBN 0-306-45247-2 (hardbound).—ISBN 0-306-45324-X (pbk.)
    1. Materials—Microscopy. 2. Transmission electron microscopy. I. Carter, C. Barry. II. Ti-
tle.
TA417.23.W56   1996   96-28435
502′.8′25—dc20   CIP

ISBN 0-306-45247-2 (Hardbound)
ISBN 0-306-45324-X (Paperback)

© 1996 Plenum Press, New York
A Division of Plenum Publishing Corporation
233 Spring Street, New York, N. Y. 10013

10 9 8 7 6 5 4 3 2

Printed in the United States of America

X-ray Spectrometry

32

CHAPTER PREVIEW

To use the X-rays generated when the electron beam strikes a TEM specimen, we have to detect them first and then identify them as coming from a particular element. This is accomplished by X-ray spectrometry, which is one way we can transform a TEM into a far more powerful instrument, called an analytical electron microscope (AEM). Currently, the only kind of X-ray spectrometer that we use in an AEM is an X-ray energy-dispersive spectrometer (XEDS), which comprises a detector interfaced to signal-processing electronics and a computer-controlled multi-channel analyzer (MCA) display. The XEDS is a complex and rather sophisticated piece of instrumentation which takes advantage of modern semiconductor technology. The principal component of the XEDS is a semiconductor detector which has the benefit of being compact enough to fit within the confined region of the TEM stage and, in one form or another, is sensitive enough to detect all the elements above Li in the periodic table.

    We start with the basic physics you need to understand how the detector works and give you a very brief overview of the processing electronics. We then describe a few simple tests you can perform to confirm that your XEDS is working correctly. It is really most important from a practical point of view that you know the limitations of your XEDS system. Therefore, we will describe these limitations in some detail, especially the unavoidable artifacts. Finally, we briefly mention the wavelength-dispersive spectrometer (WDS), which is used in bulk X-ray microanalysis. The WDS is old technology which might see a renaissance in the AEM in the future.

X-ray Spectrometry

32

32.1. X-RAY ANALYSIS: WHY BOTHER?

When characterizing a specimen in the TEM, the limitations of only using imaging should by now be obvious to you. Our eyes are accustomed to the interpretation of 3D reflected-light images. However, as we have seen in great detail in Part III, the TEM gives us two-dimensional projected images of 3D transparent specimens and you, the operator, need substantial experience in order to interpret the images correctly. For example, Figure 32.1 shows six images, taken with both light and electron microscopes (can you distinguish which images are from which kind of microscope?). The scale of the microstructures varies from nanometers to millimeters and yet the images appear very similar. Without any prior knowledge it would be almost impossible, even for an experienced microscopist, to identify the nature of these specimens simply by looking at the image.

Now if you look at Figure 32.2, you can see six X-ray spectra, one from each of the specimens in Figure 32.1. We will be discussing such spectra in detail later in this and subsequent chapters, but even with no knowledge of XEDS, you can easily see that each specimen gives a different spectrum.

> Different spectra mean that each specimen must have a different elemental composition and it is possible to obtain this information in a matter of minutes.

Armed with the elemental make-up of your specimen, any subsequent image and diffraction analysis is greatly facilitated. For your interest, the identity of each specimen is given in the caption to Figure 32.2. While Figures 32.2A–E are all from common inorganic materials, Figure 32.2F is from a cauliflower which, once you get it into the electron microscope, provides a very distinctive spectrum. The familiar morphology of this specimen, now obvious in Figure 32.1F, also accounts for the generic term "cauliflower structure" which is given to these and other similar microstructures.

> The main message you should get from this illustration is that the *combination* of imaging and spectrometry is most powerful and this combination transforms a TEM to an AEM.

Within a very short time, you can get a qualitative elemental analysis of most features in a complex microstructure, and the important features can be isolated for full quantitative analysis, which we will address in Chapter 35. In the chapters before this you will first learn something about how the XEDS detector works and the problems that arise when the detector is inserted into the column of an AEM.

32.2. BASIC OPERATIONAL MODE

To produce spectra such as those in Figure 32.2, all that you have to do is obtain a TEM or STEM image of the area you wish to analyze. In TEM mode, you then have to condense the beam down to an appropriate size for analysis. This may mean exciting the C1 lens more strongly and changing the C2 aperture and C2 lens strength. These steps may misalign the illumination system. For this reason it is recommended that you operate in STEM mode. Create your STEM image using the appropriate C1 lens setting and C2 aperture to give the desired probe dimension. It is then a simple matter to stop the scanning probe and position it on the feature you wish to analyze. Furthermore, critical software routines that you can use to check for specimen drift during your analysis can only work via a digital image of the analysis region.

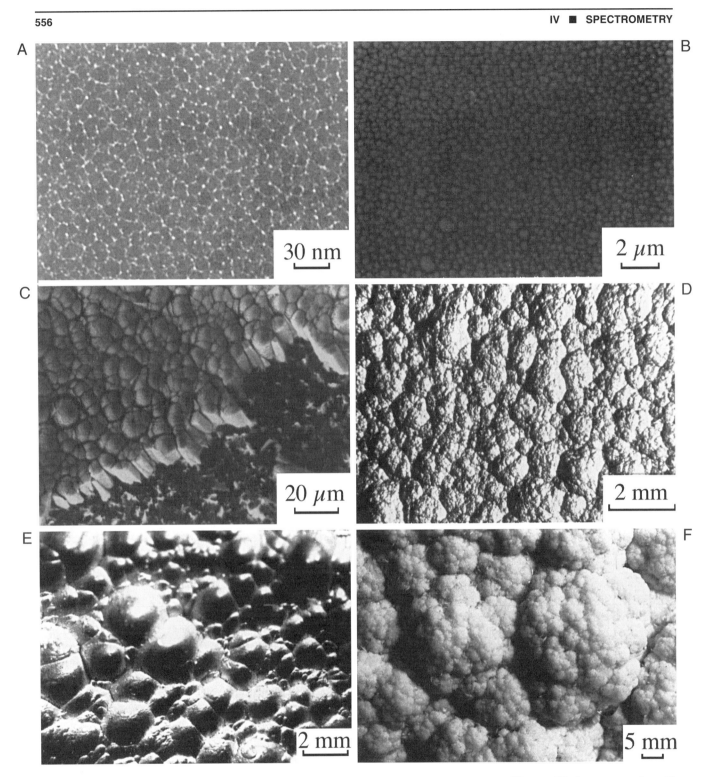

Figure 32.1. Six images of various microstructures, spanning the dimensional range from nanometers to millimeters. The images were taken with TEMs, SEMs, and light microscopes, but the characteristic structures are very similar, and it is not possible, without prior knowledge, to identify the samples.

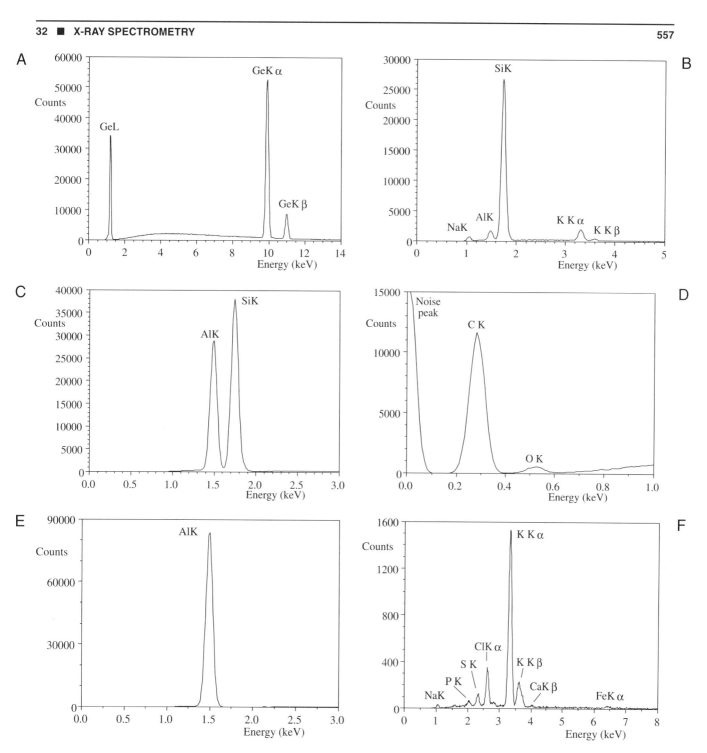

Figure 32.2. XEDS spectra from the six materials in Figure 32.1. Each spectrum is clearly different from the others, and helps to identify the samples as (A) pure Ge, (B) silica glass, (C) Al evaporated on a Si substrate, (D) pyrolitic graphite, (E) pure Al, and (F) a cauliflower.

Use STEM images to select your analysis region.
This makes it easier to move between image mode
and analysis mode.

Microanalysis should *always* be performed with your specimen in a low-background (Be) holder (see Chapter 8). The
holder should be capable of being cooled to liquid-N_2 temperature to minimize contamination, and a double-tilt version is recommended so diffraction and imaging can be
carried out simultaneously.

32.3. THE ENERGY-DISPERSIVE SPECTROMETER

The XEDS produces spectra which are plots of X-ray
counts (imprecisely termed "intensity") versus X-ray *energy*. Before we get into details, recall from back in Chapter 4 that electrons generate two kinds of X-rays. When
electrons ionize an atom, the emitted characteristic X-ray
energy is unique to the ionized atom. When electrons are
slowed by interaction with the nucleus, they produce a continuum of bremsstrahlung X-rays. The result is that, as we
have seen in Figures 1.4A, 4.6, and 32.2, the characteristic
X-rays appear as Gaussian-shaped peaks superimposed
on a background of bremsstrahlung X-rays, most clearly
visible in Figure 32.2A. Many more spectra will appear
throughout this and subsequent chapters.

The XEDS was developed in the late 1960s and, by
the mid-1970s, it was available as an option on many
TEMs and even more widespread on other electron beam
instruments, such as SEMs. This testifies to the fact that
the XEDS is really quite a remarkable instrument, embody-
ing many of the most advanced features of semiconductor
technology. It is compact, stable, robust, easy to use, and
you can quickly interpret the readout. Several books have
been devoted to XEDS and these are listed in the general
reference section. Figure 32.3A shows a schematic diagram of an XEDS system and we'll deal with each of the
major components as we go through this chapter.

The three main parts are the detector, the processing electronics, and the MCA display.

A computer controls all three parts. First, it controls
whether the detector is on or off. Ideally, we only want to
process one incoming X-ray at a time so the detector is
switched off when an X-ray signal is detected. Second, the
computer controls the processing electronics, setting the
time required to analyze the X-ray signal and assigning

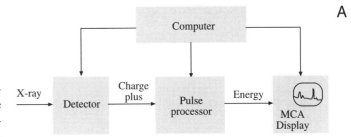

A

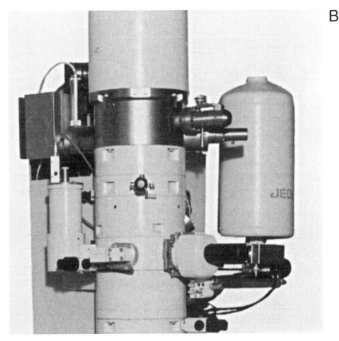

B

Figure 32.3. (A) Diagram of the XEDS system showing how the computer controls the detector, the processing electronics, and the display.
(B) An XEDS system interfaced to the stage of an AEM. All that is visible
is the large liquid-N_2 dewar attached to the side of the column.

the signal to the correct channel in the MCA. Third, the
computer software governs both the calibration of the
spectrum readout on the MCA screen and all the alpha-
numerics which tell you the conditions under which you
acquired the spectrum. Any data processing is also carried
out using the computer.

We can summarize the working of the XEDS as
follows:

■ The detector generates a charge pulse proportional to the X-ray energy.
■ This pulse is first converted to a voltage.
■ Then the signal is amplified through a field
   effect transistor (FET), isolated from other
   pulses, further amplified, then identified elec-

tronically as resulting from an X-ray of specific energy.

■ Finally, a digitized signal is stored in a channel assigned to that energy in the MCA.

The speed of this process is such that the spectrum appears to be generated in parallel with the full range of X-ray energies detected simultaneously, but the process actually involves very rapid serial processing of individual X-ray signals. Thus the XEDS both detects X-rays and separates (disperses) them into a spectrum according to their *energy*, hence the name of the spectrometer.

Figure 32.3B shows a detector interfaced to an AEM. In fact, you can't see the processing electronics or the MCA display, nor the detector itself because it sits close to the specimen within the microscope column. The most prominent feature that you can see is the liquid-N$_2$ dewar, which cools the detector.

32.4. SEMICONDUCTOR DETECTORS

The detector in an XEDS is a reverse-biased p-i-n diode. Almost all AEMs use silicon–lithium [Si(Li)] semiconductor detectors and so we will take these as our model. Later, in Section 32.4.C, we'll discuss the role of intrinsic Ge (IG) detectors, which can be useful on intermediate voltage AEMs.

32.4.A. How Does XEDS Work?

While you don't need to understand precisely how the detector works in order to use it, a basic understanding will help you optimize your system and it will also become obvious why certain experimental procedures and precautions are necessary.

When X-rays interact with a semiconductor, the primary method of energy deposition is the transfer of electrons from the valence band to the conduction band, creating an electron–hole pair. High-energy electrons lose energy in Si in a similar way as we saw in Section 4.4. The energy required for this transfer in Si is ~3.8 eV at the liquid-N$_2$ operating temperature. (This quantity is a statistical value so don't try to link it directly to the band gap.) Since characteristic X-rays typically have energies well in excess of 1 keV, thousands of electron–hole pairs can be generated by a single X-ray. The number of electrons or holes created is directly proportional to the energy of the incoming X-ray. Even though all the X-ray energy is not, in fact, converted to electron–hole pairs, enough are created for us to collect sufficient signal to distinguish most elements in the periodic table with good statistical precision. The way this is achieved is summarized in Figure 32.4, which is a schematic diagram of the Si(Li) detector.

The design of the Si(Li) detector is very similar to the semiconductor electron detectors discussed in Chapter 7.

> The electron detectors separate the electrons and holes by the internal reverse bias of a very narrow p-n junction; since X-rays penetrate matter much more easily than electrons, we need a much thicker region for the X-rays to generate electron–hole pairs and lose all their energy.

In practice, we need to have an intrinsic region between p- and n-type regions which is about 3 mm thick. So the Si should have low conductivity, with no impurity atoms to contribute electrons or holes to the charge pulse, and no defects to act as recombination sites for the electron–hole pairs. However, we still cannot make intrinsic Si on a commercial basis. It usually contains acceptor impurities and so acts as a p-type semiconductor. We compensate for the impurity effects by "filling" any recombination sites with Li, thus creating a region of intrinsic Si, hence the term Si(Li), popularly pronounced "silly." Without the Li, commercial-purity Si would suffer electrical breakdown when the bias was applied to separate the electrons and holes. The Li is introduced either by diffusion under an applied voltage (hence the term "Li-drifted" detector) or, in a more controlled fashion, by ion implantation followed by a diffusion anneal.

While many electrons and holes are generated by an X-ray, they still constitute a very small charge pulse (about 10^{-16} C), and so a negative bias of ~0.5–1 keV is ap-

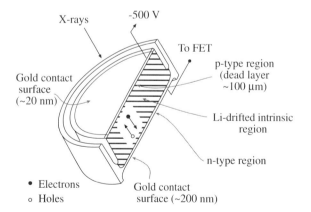

Figure 32.4. Cross section of a Si(Li) detector. The incoming X-rays generate electron–hole pairs in the intrinsic Si which are separated by an applied bias. A positive bias attracts the electrons to the rear ohmic contact after which the signal is amplified by an FET.

plied across the Si to ensure collection of most of the signal. We apply the bias between ohmic contacts, which are evaporated metal films such as Au or Ni, ~10–20 nm thick for the front face and ~200 nm at the back. This metal film also produces a p-type region at the front of the crystal; the back of the crystal is doped to produce n-type Si.

> So the whole crystal is now a p-i-n device, with relatively shallow junctions less than 200 nm deep at either side of the central 3-mm intrinsic region.

When a reverse bias is applied to the crystal (i.e., a negative charge is placed on the p-type region at the front of the detector and a positive charge on the rear), the electrons and holes are separated and a charge pulse of electrons can be measured at the rear ohmic contact. Remember that the magnitude of this pulse is proportional to the energy of the X-ray that generated the electron–hole pairs. (We could equally well measure the whole pulse, but available low-noise FETs are n-channel devices, requiring electron collection.)

The p and n regions, at either end of the detector, are usually termed "dead layers." The traditional argument for use of this term is that the Li compensation is not completely effective so most of the electron–hole pairs generated in these end regions recombine, and contribute nothing to the charge pulse. Recently, however, Joy (1995) has shown that the dead layer can be explained if the diffusion length of the charge-carrying electrons exceeds the distance they travel under the drift field, in which case they are not "dead" but they will not be gathered at the surface electrode and contribute to the charge pulse. However, we'll still use the inaccurate term "dead layer" because it is so common. In practice it is the layer at the entrance surface of the detector that is most important since the X-rays must traverse it to be detected, and we will refer to this as *the* dead layer. This dead layer affects the spectrum, particularly when you are studying peaks from the low-Z elements (McCarthy 1995).

> The p and n regions are called "dead layers" and the intrinsic region in between is referred to as the "active layer."

Why do we have to cool the detector with liquid N_2? Well, if the detector were at room temperature, three highly undesirable effects would occur:

■ Thermal energy would activate electron–hole pairs, giving a noise level that would swamp the X-ray signals we want to detect.

■ The Li atoms would diffuse under the applied bias, destroying the intrinsic properties of the detector.

■ The noise level in the FET would mask signals from low-energy X-rays.

For these reasons we cool the detector and the FET with liquid N_2, necessitating the characteristic dewar mentioned above (see Figure 32.3B). The FET gets to a temperature of about 140 K and the detector surface is at about 90 K.

Cooling the detector and the FET brings some undesirable consequences, which we have to accommodate. The minor irritations are that we have to regularly monitor and fill up the liquid-N_2 dewar. The more severe consequence of the cooling is that both hydrocarbons and ice from the microscope environment can condense on the cold detector surface, causing absorption of lower-energy X-rays. There are two obvious solutions to this problem. Either we can isolate the detector from the microscope vacuum, or we can remove hydrocarbons and water vapor from the microscope. The latter is the more desirable solution but the former is far easier. So, detectors are sealed in a prepumped tube with a "window" to allow X-rays through into the detector.

> You have a choice of three different kinds of detector: those with a Be window, those with an ultrathin window, and those without a protective window.

Let's examine the pros and cons of each detector window; a good review has been given by Lund (1995).

32.4.B. Different Kinds of Windows

Beryllium window detectors use a thin sheet (nominally 7 μm) of beryllium which is transparent to most X-rays, and can withstand atmospheric pressure when the stage is vented to air. (In fact, 7 μm Be is expensive ($3M/pound!), rare, and slightly porous, so thicker sheet (> ~ 12 μm) is more commonly used.) Production of such thin Be sheet is a remarkable metallurgical achievement, but the window is still too thick to permit passage of all characteristic X-rays; any that have energy less than ~1 keV are strongly absorbed. Therefore, we cannot detect K_α X-rays from elements below about Na ($Z = 11$) in the periodic table. The Be window prevents microanalysis of the lighter elements such as B, C, N, and O, which are important in materials science (and also in other disciplines that use the AEM, such as the biological and geological sciences). Other factors, such as the low fluorescence yield and absorption within the specimen, make light-element X-ray microanalysis somewhat of a challenge, and EELS is often preferable.

Ultrathin window (UTW) detectors have windows that are less absorbent than Be; usually these are made from very thin (<100 nm) films of polymer, diamond, boron nitride, or silicon nitride, all of which are capable of withstanding atmospheric pressure while still transmitting 192-eV boron K_α X-rays. Early UTWs were very thin polymer membranes, such as parylene. Unfortunately, these were unable to withstand atmospheric pressure; they had to be withdrawn behind a vacuum isolation valve whenever a specimen was exchanged or the column was vented to air. Newer, composite Al/polymer UTWs or very thin diamond or BN windows, sometimes termed "atmospheric thin windows" (ATWs), are really the only sensible option. You should remember that different window materials absorb the light-element X-rays differently, so you need to know the characteristics of the window in the particular system you are using. For example, carbon-containing windows absorb N K_α X-rays very strongly.

Windowless detectors were first tried in the early 1970s, but microscope vacuums were relatively poor, resulting in rapid hydrocarbon and/or ice contamination of the detector surface. You should only use windowless detectors in a UHV AEM. Take great care to eliminate hydrocarbons from your specimen and keep the partial pressure of water vapor below ~10^{-8} Pa by efficient pumping. The best performance by a windowless system is the detection of Be (110 eV) K_α X-rays as shown in Figure 32.5, which is a remarkable feat of electronics technology.

> You may recall that it takes ~3.8 eV to generate an electron–hole pair in Si, so a Be K_α X-ray will create at most 29 electron–hole pairs, giving a charge pulse of ~5×10^{-18} C !

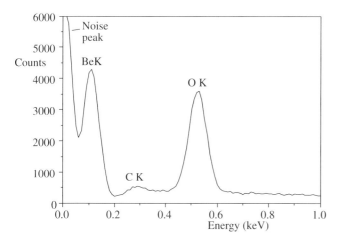

Figure 32.5. XEDS spectrum showing the detection of Be in an oxidized Be foil in an SEM at 10 keV. The Be K_α line is not quite resolved from the noise peak.

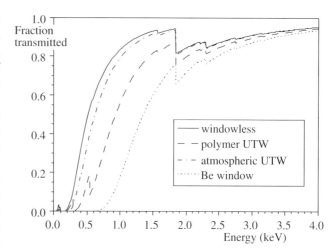

Figure 32.6. Low-energy efficiency calculated for a windowless detector, UTW detector (1 μm Mylar coated with 20 nm of Al), an ATW detector and a 13-μm Be window detector. Note that the efficiency is measured in terms of the fraction of X-rays transmitted by the window.

The relative performance of the various types of Si(Li) detector windows is summarized in Figure 32.6. Here we plot the detector efficiency as a function of the energy of the incoming X-ray. You can clearly see the rapid drop in efficiency at the low-energy end and the increased efficiency of UTW and windowless detectors. In fact, the Si(Li) detector absorbs X-rays with almost 100% efficiency over the energy range from about 2 to 20 keV, as shown in Figure 32.7. Within this energy range you will find X-rays from all the elements in the periodic table

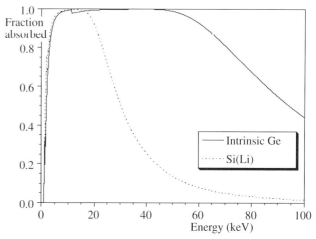

Figure 32.7. High-energy efficiency up to 100-keV X-ray energy calculated for Si(Li) and IG detectors, assuming a detector thickness of 3 mm in each case. Note the large effect of the Ge absorption edge at about 11 keV. In contrast to Figure 32.6, the efficiency in this case is measured by the fraction of X-rays absorbed within the detector.

above phosphorus. This uniform high efficiency is a major advantage of the XEDS detector.

32.4.C. Intrinsic Germanium Detectors

You can see in Figure 32.7 that Si(Li) detectors show a drop in efficiency above ~20 keV. This is because X-rays with such high energy can pass through the detector without depositing their energy by creating electron–hole pairs. This effect limits the use of Si(Li) detectors in intermediate voltage AEMs because at 300–400 keV we can generate K_α X-rays from all the high-atomic-number elements; Pb K_α X-rays at 75 keV are easily excited by 300-keV electrons. As we'll see in Chapter 35, there are certain advantages to using the K lines rather than the lower-energy L or M lines for quantification; with a Si(Li) detector the K lines from elements above silver (Z = 47) are barely detectable. The answer to this problem is to use a Ge detector, which more strongly absorbs high-energy X-rays (Sareen 1995).

We can manufacture Ge of higher purity than Si, and therefore Li compensation is not needed to produce a large intrinsic region; clearly this is a major advantage. Like the Si(Li) detector, the intrinsic Ge (IG) or high-purity Ge (HPGe) detector can have a Be window, a UTW/ATW, or no window; in any form it has some advantages over Si(Li). The detector is more robust and it can be warmed up repeatedly, which, as we'll see, sometimes solves certain problems.

> The intense doses of high-energy electrons or X-rays which can easily be generated in an AEM (e.g., when the beam hits a grid bar) can destroy the Li compensation in a Si(Li) detector, but there is no such problem in an IG crystal.

Furthermore, the intrinsic region can easily be made ~5 mm thick, which results in 100% efficient detection of Pb K_α X-rays at ~ 75 keV. Figure 32.7 compares the efficiency of Si(Li) and IG detectors up to 100 keV and Figure 32.8 shows detection of Pb $K_{\alpha1}$ and $K_{\alpha2}$ lines generated from a lead glass specimen at 200 kV.

There is an even more fundamental advantage to IG detectors. Since it takes only ~2.9 eV of energy to create an electron–hole pair in Ge, compared with 3.8 eV in Si, a given X-ray produces more electron–hole pairs, and so the energy resolution and signal to noise are better. However, as you may have guessed, there are some difficulties in using IG detectors. The high-energy K lines, for which these detectors are ideally suited, have very small ionization cross sections when using 300–400 keV electrons, and so the spectral intensities are rather low, as you can see in Figure 32.8. A minor drawback is that IG detectors have to be

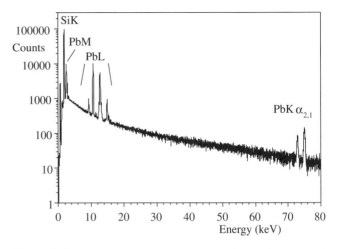

Figure 32.8. High-energy spectrum from lead silicate glass analyzed in a 200-kV AEM with an IG detector. Note the logarithmic scale for the counts which masks somewhat the very low intensity of the K lines compared with the L and M families. Note also that the $K_{\alpha1}$ and $K_{\alpha2}$ lines are clearly resolved.

cooled 25 K lower than Si(Li) to give the same leakage current, and they invariably need an Al-coated UTW since they are more sensitive to infrared radiation than Si(Li) detectors. This UTW reduces the maximum collection angle of IG detectors (see Section 33.2). Si(Li) detectors are easier to manufacture and are more reliable. They have a long history of dependable operation and a large number are already in use. IG detectors should eventually become more widely accepted as users install new systems.

IG detectors have been used since ~1970 by nuclear physicists to detect MeV radiation, but the first one was not installed on an AEM until 1986. One reason for this slow transfer of technology was that the high-energy K lines were very inefficiently excited in the lower voltage AEMs available at the time. Also, the low-energy performance of IG detectors was very poor in early detectors due mainly to a thick dead layer, which resulted in very non-Gaussian peak shapes. This problem has now been overcome and Gaussian characteristic peaks can be generated across the full energy range of the spectrum.

> In fact, UTW IG detectors are capable of detecting X-rays from boron to uranium, although the low-energy spectrum is still a little better in a Si(Li) system.

It is arguable that all intermediate voltage AEMs should be equipped with two detectors, an IG and a UTW Si(Li), to give the most efficient X-ray detection across the widest possible elemental range. Look ahead to Table 32.1 for a comparison of the two kinds of detector.

32.5. PULSE PROCESSING AND DEAD TIME

The electronic components attached to the detector convert the charge pulse created by the incoming X-ray into a voltage pulse, which can be stored in the appropriate energy channel of the MCA. The pulse-processing electronics must maintain good energy resolution across the spectrum without peak shift or distortion, even at high counting rates. To accomplish this, all the electronic components beyond the detector crystal must have low-noise characteristics and must employ some means of handling pulses that arrive in rapid succession. Currently, this whole process relies on analog pulse processing, but it is likely that, in the near future, many of the problems we'll now describe will be solved by digital techniques (Mott and Friel 1995).

Let's consider first of all what happens if a single isolated X-ray enters the detector and creates a pulse of electrons at the back of the Si(Li) crystal.

- ■ The charge pulse enters the FET, which acts as a preamplifier and converts the charge into a voltage pulse.
- ■ This voltage pulse is amplified several thousand times by a pulse processor, and shaped so that an analog-to-digital converter can recognize the pulse as coming from an X-ray of specific energy. (The XEDS doesn't do a very good job of accurate energy assessment.)
- ■ The computer assigns it to the appropriate channel in the MCA display.

The accumulation of pulses or counts entering each energy channel at various rates produces a histogram of counts versus energy that is a digital representation of the X-ray spectrum. The MCA display offers multiples of 1024 channels in which to display the spectrum, and various energy ranges can be assigned to these channels. For example, 10, 20, or 40 keV full horizontal scales can be used (or even 80 keV for an IG detector on an intermediate voltage AEM). The display resolution chosen depends on the number of channels available.

> A typical energy range that you might select for a Si(Li) detector is 20 keV, and in 2048 channels this gives you a display resolution of 10 eV per channel.

You should keep the display resolution at about 10 eV per channel. A smaller value ties up a lot of memory and often you can't display the whole spectrum at once. A larger value gives you only a few channels for each characteristic peak. For an IG detector, more channels (at least 4096) are needed to display the complete spectrum up to 80 or 100 keV, but the resolution of the MCA display is usually poorer, ~20 eV/channel.

Details of the pulse processing electronics are not important except for two variables over which you have control. These are the time constant and the dead time. The *time constant* (τ) is the time (~10–50 μs) allowed for the pulse processor to evaluate the magnitude of the pulse. If you select a longer τ, the system is better able to assign an energy to the incoming pulse, but fewer counts can be processed in a given analysis time. You have a choice of τ given by the manufacturer:

- ■ The shortest τ (typically a few μs) will allow you to process more counts per second (cps) but with a greater error in the assignment of a specific energy to the pulse, and so the energy resolution (see Section 32.6 below) will be poorer.
- ■ A longer τ (up to about 50 μs) will give better resolution but the count rate will be lower.

> You can't have a high count rate and good resolution, so for most routine thin-foil analyses you should maximize the count rate (shortest τ), unless there is a specific reason why you want to get the best possible energy resolution (longest τ). This recommendation is based on a detailed argument presented by Statham (1995).

Now in reality there are many X-rays entering the detector, but because of the speed of modern electronics the system can usually discriminate between the arrival of two almost simultaneous X-rays. The details of the electronics can be found, e.g., in Goldstein *et al.* (1992). When the electronic circuitry detects the arrival of a pulse, it takes less than a microsecond before the detector is effectively switched off for the period of time called the *dead time* while the pulse processor analyzes that pulse. The dead time is clearly closely related to τ, which is so small that you should expect your detector system to process up to 10,000 cps quite easily. The dead time will increase as more X-rays try to enter the detector, which closes down more often. The dead time can be defined in several ways. Take the ratio of the output count rate (R_{out}) to the input count rate (R_{in}), which you can usually measure. Then we can say

$$\text{Dead time in } \% = \left(1 - \frac{R_{out}}{R_{in}}\right) \times 100\% \qquad [32.1a]$$

An alternative definition is

$$\text{Dead time in \%} = \frac{(\text{clock time} - \text{live time})}{\text{live time}} \times 100\% \quad [32.1b]$$

Put another way, this equation says that if you ask the computer to collect a spectrum for a "live time" of 100 s, this means that the detector must be "live" and receiving X-rays for this amount of time. If it is actually dead for 20 s while it is processing the X-rays, then the dead time will be 20%, and it will take 120 s of "clock time" to accumulate a spectrum. As the input count rate increases, the output count rate will drop and the clock time will increase accordingly. Dead times in excess of 50–60% (or as little as 30% in old systems) mean that the detector is being swamped with X-rays and collection becomes increasingly inefficient, and it is better to turn down the beam current or move to a thinner area of the specimen to lower the count rate.

32.6. RESOLUTION OF THE DETECTOR

We can define the energy resolution R of the detector as follows

$$R^2 = P^2 + I^2 + X^2 \quad [32.2]$$

The term P is a measure of the quality of the associated electronics. It is defined as the full width at half maximum (FWHM) of a randomized electronic pulse generator. X is the FWHM equivalent attributable to detector leakage current and incomplete charge collection (see below). I is the intrinsic line width of the detector which is controlled by fluctuations in the numbers of electron–hole pairs created by a given X-ray and is given by

$$I = 2.35 \left(F \varepsilon E \right)^{\frac{1}{2}} \quad [32.3]$$

Here, F is the Fano factor of the distribution of X-ray counts from Poisson statistics, ε is the energy to create an electron–hole pair in the detector, and E is the energy of the X-ray line. Because of these two factors, the experimental resolution can only be defined under specific analysis conditions.

The IEEE standard for R is the FWHM of the Mn K_α peak, generated (off the microscope) by an Fe^{55} source which produces 10^3 cps with an 8-μs pulse-processor time constant.

Rather than using radioactive Fe^{55}, we recommend measuring the detector resolution on the AEM column!

Now, since Mn is not a common sample to have lying around, you will find it useful to keep a thin Cr-film specimen to check the resolution when the detector is on the column. An evaporated Cr film about 100 nm thick supported on a carbon film and a Cu grid is ideal. Because Cr is next to Mn in the periodic table, the resolution of the K_α peak will be just a few eV less than that of Mn. The Cr is also stable with a resilient thin oxide film; it doesn't degrade in the electron beam. As we'll see later, these Cr films are very useful for other calibration checks and performance criteria (Zemyan and Williams 1994).

Many XEDS computer systems have an internal software routine which measures the resolution. Alternatively, you can gather a Mn or Cr peak, and select a window encompassing the peaks from the channels on both sides of the peak that contain half the maximum counts in the central channel, as shown in Figure 32.9.

Typically, Si(Li) detectors have a resolution of ~140 eV at Mn K_α with the best being 127 eV. The best reported IG resolution is 114 eV.

Because the value of ε is lower for Ge (2.9 eV) than for Si (3.8 eV), IG detectors have higher resolution than Si(Li). The resolution is also a function of the area of the detector, and the values given relate to the performance of 10-mm² detectors. The 30-mm² detectors which are usually installed on AEMs have resolutions about 5 eV worse than the figures just mentioned. However, you should also be aware that when R is measured on the microscope, there

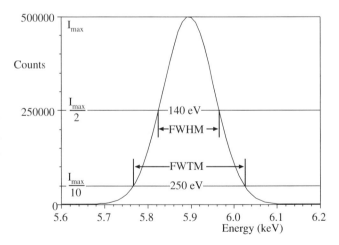

Figure 32.9. Measurement of the energy resolution of an XEDS detector by determining the number of channels that encompass the FWHM of the Mn K_α peak. The number of channels multiplied by the eV per channel gives the resolution, which typically should be about 130–140 eV. We can measure the FWTM also to give an indication of the degree of the incomplete charge collection which distorts the low-energy side of the peak. The FWTM should be ~1.82 times the FWHM.

may be a further degradation in resolution. It is rare to find a 30-mm^2 Si(Li) detector delivering a resolution much better than 140 eV on the AEM column, even though the quoted values are typically ~10 eV lower.

> Remember also that there is always a trade-off between resolution and count rate, unless digital pulse processing is used.

How close are XEDS detectors to their theoretical resolution limit? If we assume that there is no leakage and the electronics produced no noise, then $P = X = 0$ in equation 32.2, so $R = I$. For Si, $F = 0.1$, $\varepsilon = 3.8$ eV, and the Mn K_α line occurs at 5.9 keV, which gives $R = 111$ eV. So it seems that there is not much more room for improvement. The resolution of XEDS detectors won't approach that of crystal spectrometers, which is 5 to 10 eV, although, because of the dependence of I on the X-ray energy, light-element K lines have widths <<100 eV.

32.7. WHAT YOU SHOULD KNOW ABOUT YOUR XEDS

There are several fundamental parameters of both Si(Li) and IG XEDS systems which you can specify, measure, and monitor to ensure that your system is performing acceptably. Many of these tests are standard procedures (e.g., see the XEDS laboratories in Lyman *et al.* 1990) and have been summarized by Zemyan and Williams (1995). In an SEM, which is relatively well behaved, Si(Li) detectors have been known to last ten years or more before requiring service or replacement. In contrast, an AEM (particularly the higher-voltage variety) is a hostile environment and the life of a detector is often less than three years. For this reason, most detectors are equipped with protective shutters (see Section 33.3).

> It is particularly important to monitor the detector performance on your AEM, in order that quantitative analyses you make at very different times may be compared in a valid manner.

You need to know both the operating specifications for your own system and how to measure them. We can break these specifications down into detector variables and signal-processing variables.

32.7.A. Detector Variables

The detector resolution that we just defined may degrade for a variety of reasons. Two are particularly common:

■ Damage to the intrinsic region by high-energy fluxes of radiation.

■ Bubbling in the liquid-N$_2$ dewar due to ice crystals building up.

You can help to minimize the ice build-up by filtering the liquid N$_2$ before putting it into the dewar. Never re-cycle liquid N$_2$ into the dewar; use it elsewhere. If the nitrogen in the dewar is bubbling, you should consider warming up the detector, but do it *after consultation with the manufacturer, and without the applied bias*. (Think what happens to the Li otherwise.) Emptying out the dewar, filling it with hot water, and then drying it with a hair dryer will often solve the problem. However, after several such cycles, you may find that the detector resolution doesn't return to acceptable levels (the window seal may develop a leak due to the repeated thermal oscillations). When this happens, it is necessary to return the detector to the manufacturer to have it repaired.

Incomplete Charge Collection (ICC). Because of the inevitable presence of the dead layer, the X-ray peak will not be represented by a perfect Gaussian shape when displayed after processing. Usually, the peak will have a low-energy tail, because some energy will be deposited in the dead layer and will not create detectable electron–hole pairs. You can measure this ICC effect from the ratio of the full width at tenth maximum (FWTM) to the FWHM of the displayed peak, as shown schematically in Figure 32.9.

> The ideal value for FWTM/FWHM is 1.82 (Mn K_α or Cr K_α), but this value will be larger for X-ray peaks from the lighter elements, which are more strongly absorbed by the detector.

In Si(Li) detectors, the phosphorus K_α peak shows the worst ICC effects because this X-ray fluoresces Si very efficiently. ICC will also occur if a detector has a large number of recombination sites arising, for example, through damage from a high flux of backscattered electrons. The crystal defects that act as recombination sites may be annealed out by warming the detector, as we just described. IG detectors used to show worse ICC effects than Si(Li) detectors, but this is no longer the case. Now, an IG detector should meet the same FWTM/FWHM ratio criterion as a Si(Li) detector. If the ratio is higher than 2 for the Cr peak, there is something seriously wrong with the detector and you should have it replaced.

Detector Contamination. Over a period of time, ice and/or hydrocarbons will eventually build up on the cold detector surface or on the window. If ice or hydrocarbon contamination does occur, it will reduce the efficiency with which we detect low-energy X-rays. While this is most likely to happen for a windowless detector, Be and UTW systems also suffer the same problem because of residual water vapor in the detector vacuum or because the window may be slightly porous. In all cases the problem is insidious, because the effects may develop over many months and you will not notice the degradation of the spectrum until differences in light-element quantification are apparent from the same specimen analyzed at different times. Therefore, you should regularly monitor the quality of the low-energy spectrum. The ratio of the NiK_α/NiL_α has been used (Michael 1995), but you can also use the CrK_α/CrL_α intensity ratio on the same evaporated film of pure Cr used for resolution, as long as there is no significant oxide film, since the $O\ K_\alpha$ line overlaps the $Cr\ L_\alpha$ line.

> The NiK_α/NiL_α ratio will rise with time if contamination or ice is building up on the detector and selectively absorbing the lower-energy L line.

The K/L ratio will differ for different detector dead layers, for different UTWs or ATWs, and for different specimen thicknesses. So we can't define an accepted figure of merit. The best you can do is to measure the ratio immediately after installing a new (or repaired) detector and be aware that as the ratio increases, then any quantification involving similar low-energy X-ray lines will become increasingly unreliable. When the ratio increases to what you deem to be an unacceptably high value, then you must remove the ice/contamination. Automatic *in situ* heating devices which raise the detector temperature sufficiently to sublime the ice make this process routine. If your detector doesn't have such a device, then you should warm up the detector, as we described above.

In summary, you should measure and continually monitor changes in:

■ *Your detector resolution* at the Mn or CrK_α line [typically 140–150 eV for Si(Li) and 120–130 eV for IG].
■ *The ICC* defined by the FWTM/FWHM ratio of the $Cr\ K_\alpha$ line (ideally 1.82).
■ *The ice build-up* reflected in the Ni (or Cr) K_α/L_α ratio.

If any of these figures of merit get significantly larger than the accepted values, then you should have your detector serviced by the manufacturer. Warming up the detector may cure some or all of the detector problems, but it could be an expensive procedure if you get it wrong. So only do it with the bias off and after consultation with the manufacturer.

So, you must be very careful with your XEDS:

■ *Do not* generate high fluxes of X-rays or backscattered electrons unless your detector is shuttered.
■ *Do not* warm up the detector with the bias applied and without consulting the manufacturer.
■ *Do not* use unfiltered or re-cycled liquid N_2.

32.7.B. Processing Variables

Other ways to monitor the performance of the XEDS system relate to the processing of the detector output signal. There are three things you have to check to make sure the pulse processing electronics are working properly:

■ First, check the calibration of the energy range of the spectrum.
■ Second, check the dead-time correction circuitry.
■ Third, check the maximum output count rate.

The energy calibration should not change significantly from day to day, unless you change the range or the time constant. Electronic circuit stability has improved to a level where these three checks need only be done a few times a year (or if the detector has been repaired or a new detector installed).

Calibration of the energy display range. This process is quite simple; collect a spectrum from a material which generates a pair of K X-ray lines separated by about the width of the display range (e.g., Al-Cu for 0–10 keV). Some systems use an internal electronic strobe to define zero, and in this case you only need a specimen with one K line at high energy. Having gathered a spectrum, see if the computer markers are correctly positioned at the peak centroid (e.g., Al K_α at 1.54 keV and Cu K_α at 8.04 keV). If the peak and marker are more than 1 channel (10 eV) apart, then you should re-calibrate your display using the software routine supplied by the manufacturer.

Checking the dead-time correction circuit. If the dead-time correction circuitry is working properly, the pulse processor will give a linear increase in output counts as the input counts increase, for a fixed live time.

■ Choose a specimen of a pure element, say our favorite Cr foil, which we know will give a strong K_α peak.
■ Choose a live time, say 50 s, and a beam current to give a dead-time readout of about 10%.

■ Measure the total Cr K_α counts that accumulate in 50 s.

■ Then repeat the experiment with higher input count rates.

To increase the count rate, increase the beam current by choosing a larger-diameter beam or larger C2 aperture. The dead time should increase as the input count rate goes up, but the live time will remain at the chosen value. If you plot the number of output counts against the beam current, measured with a Faraday cup, or a calibrated exposure meter reading, then it should be linear, as shown in Figure 32.10. But you will see that it will take increasingly longer clock times to attain the preset live time. If you don't have a Faraday cup, you can use the input count rate as a measure of the current; remember that the Faraday cup is useful for many other functions, such as characterizing the performance of the electron source, as we saw in Chapter 5.

Determination of the maximum output count rate. Again the procedure is simple:

■ Gather a spectrum for a fixed clock time, say 10 s, with a given dead time, say 10%.

■ Increase the dead time by increasing the beam current, C2 aperture size, or specimen thickness.

■ See how many counts accumulate in the Cr K_α peak.

The number of counts should rise to a maximum and then drop off, because beyond a certain dead time, which depends on the system electronics, the detector will be closed more than it is open and so the counts in a given clock time will decrease. In Figure 32.11, this maximum is at about 60%, typical of modern systems, although in older XEDS units this peak can occur at as little as 30% dead time. You

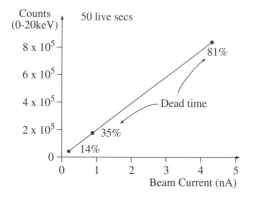

Figure 32.10.   A plot of the output counts in a fixed live time as a function of increasing beam current showing good linear behavior over a range of dead times, implying that the dead-time correction circuitry is operating correctly.

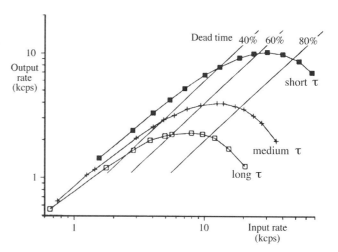

Figure 32.11.   The output count rate in a given clock time as a function of dead time. The maximum processing efficiency is reached at about 60% dead time. It is very inefficient to use the system above the maximum output rate. Increasing the time constant results in fewer counts being processed and a drop in the output count rate.

can repeat this experiment for different time constants, τ, and the counts should increase as τ is lowered (at the expense of energy resolution), as also shown in Figure 32.11. Clearly, if you operate at the maximum in such a curve (if you can generate enough input counts) then you will be getting the maximum possible information from your specimen. As we've already said, it is generally better to have more counts than to have the best energy resolution, so select the shortest τ unless you have a peak overlap problem.

While it is rare that a good thin foil produces enough X-ray counts to overload modern detector electronics, there are situations (e.g., maximizing analytical sensitivity) when it's desirable to generate as many counts as possible. Under these circumstances, use of thicker specimens and high beam currents may produce too many counts for conventional analog processing systems. As shown in Figure 32.12A, digital processing permits a higher throughput over a continuous range of energy resolution than the fixed ranges available from specific (in this case, six) time constants. Even more dramatic, as shown in Figure 32.12B, megahertz-rate beam blanking, which deflects the beam off the specimen as soon as the XEDS detects an incoming pulse, permits a remarkable increase in throughput (Lyman *et al.* 1994).

If your specimen is too thin, it might not be possible to generate sufficient X-ray counts to reach dead times in excess of 50%, so the curve may not reach a maximum, particularly if τ is very short. In this case, just use a thicker specimen.

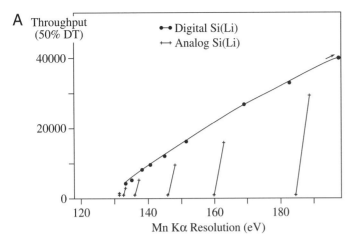

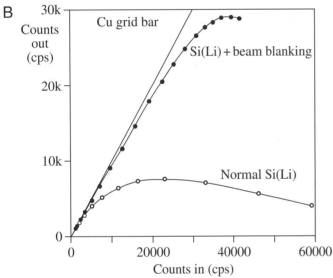

Figure 32.12. (A) Digital pulse processing gives a continuous range of X-ray throughput at 50% dead time, compared with a set of fixed throughput ranges for specific analog processing time constants. (B) Megahertz beam blanking results in a four times improvement in X-ray throughput compared to processing without beam blanking.

In summary you should occasionally:

■ Check the energy calibration of the MCA display.
■ Check the dead-time circuitry by the linearity of the output count rate versus beam current.
■ Check the counts in a fixed clock time as a function of beam current to determine the maximum output count rate.

32.7.C. Artifacts Common to XEDS Systems

The XEDS system introduces its own artifacts into the spectrum. Fortunately, we understand all these artifacts,

but they still occasionally mislead the unwary operator; see the review by Newbury (1995). We can separate the artifacts into two groups:

■ Signal-detection artifacts. Examples are "escape" peaks and "internal fluorescence" peaks.
■ Signal-processing artifacts. One example is "sum" peaks.

Escape peak. Because the detector is not a perfect "sink" for all the X-ray energy, it is possible that a small fraction of the energy is lost and not transformed into electron–hole pairs. The easiest way for this to happen is if the incoming photon of energy E fluoresces a Si K_α X-ray (energy 1.74 keV) which escapes from the intrinsic region of the detector. The detector then registers an energy of $(E-1.74)$ keV. An example is shown in Figure 32.13.

> Si "escape peaks" appear in the spectrum 1.74 keV below the true characteristic peak position.

The magnitude of the escape peak depends on the design of the detector and the energy of the fluorescing X-ray. The most efficient X-ray to fluoresce Si K_α X-rays is the P K_α, but in a well-designed detector even the P escape peak will only amount to < 2% of the P K_α intensity. This fact explains why you can only see escape peaks if there are major characteristic peaks in the spectrum. More escape peaks occur in IG spectra because we can fluoresce both Ge K_α (9.89 keV) and L_α (1.19 keV) characteristic X-rays in the detector. Each of these can cause corresponding escape peaks, but the L_α has much less chance of escaping.

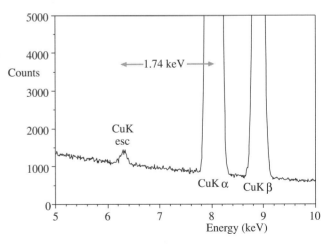

Figure 32.13. The escape peak in a spectrum from pure Cu, 1.74 keV below the Cu K_α peak. The intense K_α peak is truncated because it is ~50–100 times more intense than the escape peak.

The quantitative analysis software should be able to recognize any escape peak in the spectrum, remove it, and add the intensity back into the characteristic peak where it belongs. Because the escape peak intensity is so small it is rarely a problem.

The internal-fluorescence peak. This is a characteristic peak from the Si or Ge in the detector dead layer. Incoming photons can fluoresce atoms in the dead layer and the resulting Si K_α or Ge K/L X-rays enter the intrinsic region of the detector, which cannot distinguish their source and therefore registers a small peak in the spectrum. As detector design has improved and dead layers have decreased in thickness, the internal-fluorescence peak artifact has shrunk. However, it has not yet disappeared entirely.

> A small Si K_α peak will occur in all spectra from Si(Li) detectors for long counting times.

Obviously you must be wary when looking for small amounts of Si in a specimen, because you'll always find it! Depending on the detector, particularly the dead-layer thickness, the Si signal has an intensity corresponding to about 0.1% to 1% of the composition of the specimen (see Figure 32.14), so again it is not a major problem. Similar effects are observed in IG detectors also. The Au absorption edge from the ohmic contact layer at the front end of the detector is sometimes detectable as a small disturbance in the bremsstrahlung intensity, around 1 keV, but the effect on microanalysis is negligible.

Sum peak. As we described earlier, the processing electronics are designed to switch off the detector while each pulse is analyzed and assigned to the correct energy channel. The sum peak arises when the electronics are not

fast enough. We can identify the conditions where this is likely to occur:

- The input count rate is high.
- The dead times are in excess of about 60%.
- There are major characteristic peaks in the spectrum.

The system simply cannot be perfect. Occasionally two photons will enter the detector at almost exactly the same time. The analyzer then registers an energy corresponding to the sum of the two photons. Since this event is most likely for the X-ray giving the major peak, a sum peak (sometimes called a coincidence or double peak) appears at twice the energy of the major peak, as shown in Figure 32.15.

> The sum peak should be invisible if you maintain a reasonable input count rate, typically < 10,000 cps, which should give a dead time of < 60%.

In an AEM, you can't usually generate such high count rates unless your specimen is very thick. As always, you should at least be aware of the danger. For example, the Ar K_α energy is almost exactly twice the Al K_α energy. In the past the sum peak has led some researchers to report argon being present in aluminum specimens when it wasn't, and others to ignore argon which actually was present in ion-milled specimens! As detector electronics have improved, the sum peak has ceased to cause significant problems, except for intense low-energy peaks below ~1.25 keV, e.g., Mg K_α, where the residual noise in the electronic circuitry interferes with the pile-up rejection. So

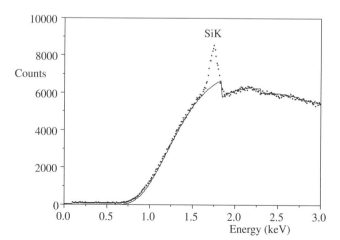

Figure 32.14. The Si internal fluorescence peak in a spectrum from pure C obtained with a Si(Li) detector. The ideal spectrum is fitted as a continuous line which exhibits the Si K absorption edge only.

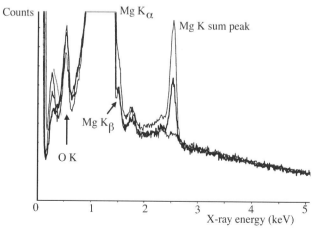

Figure 32.15. The Mg sum (coincidence) peak at various dead times; upper trace 70% dead time, middle trace 47%, lower trace 14%. The artifact is absent at 14% dead time.

if you're analyzing lighter elements than Mg, take care to use low input count rates. Reducing the dead time to 10–20% should remove even the Mg K_α sum peak, as shown in Figure 32.15.

Much of what we have just discussed can be observed experimentally on the AEM. But it is often just as instructive, and certainly easier, to simulate the spectra. To this end, we strongly advise you to purchase the simulation software Desktop Spectrum Analyzer (DTSA) from NIST, which is listed in the recommended software in Section 1.5. This software permits realistic simulation of XEDS spectra in TEMs (and SEMs) and introduces you to all the aspects of spectral processing, artifacts, modeling, etc., that are discussed in the next four chapters.

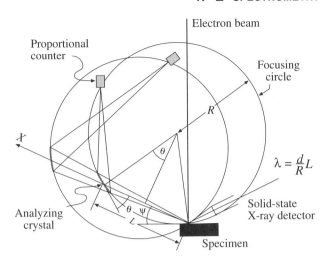

Figure 32.16.   Schematic diagram of a WDS system showing how the specimen, crystal, and detector are constrained to move on the focusing circle, radius *R*, such that the specimen–crystal distance *L* is directly proportional to the X-ray wavelength.

32.8. WAVELENGTH-DISPERSIVE SPECTROMETERS

Before the invention of the XEDS, the wavelength-dispersive spectrometer (WDS), or crystal spectrometer, was widely used. The WDS uses one or more diffracting crystals of known interplanar spacing, as devised by W. H. Bragg in 1913. Bragg's Law ($n\lambda = 2d\sin\theta$), which we've already come across in Part II, also describes the dispersion of X-rays of a given wavelength λ through different scattering angles, 2θ. We accomplish this dispersion by placing a single crystal of known interplanar spacing (d) at the center of a focusing circle which has the X-ray source (the specimen) and the X-ray detector on the circumference, as shown in Figure 32.16. The detector is usually a gas-flow proportional counter, but there's no reason why it couldn't be a Si(Li) or Ge semiconductor detector. In fact, these detectors may see more use as the need for better vacuums increases.

The mechanical motions of the crystal and detector are coupled such that the detector always makes an angle θ with the crystal surface while it moves an angular amount 2θ as the crystal rotates through θ. By scanning the spectrometer, a limited range of X-ray wavelengths of about the same dimension as the d-value of the analyzing crystal can be detected. For example, diffraction from the (200) planes of a LiF crystal covers an energy range of 3.5–12.5 keV (0.35–0.1 nm) for a scanning range of $\theta = 15$–65°. To detect X-rays outside this energy range, another crystal of different d-value must be employed.

As we'll discuss in Chapter 35, the forerunner of the AEM was the electron microscope microanalyzer (EMMA), which used a WDS. However, the WDS was a large and inefficient addition to the microscope, and never attained general acceptance by transmission microscopists.

There are two major drawbacks to the WDS compared with the XEDS:

■ The crystal has to be moved to a precise angle where it collects only a tiny fraction of the total number of X-rays coming from the specimen, whereas the XEDS detector can be placed almost anywhere in the TEM stage above the specimen and subtends a relatively large solid angle at the specimen.

■ The WDS collects a single wavelength at a given time while the XEDS detects X-rays of a large range of energies. WDS is a serial collector; XEDS is effectively a parallel collector.

The geometrical advantage in the collection efficiency of XEDS, combined with the ability to detect X-rays simultaneously over a wide energy range without the mechanical motion required of the WDS, accounts for the present dominance of XEDS systems in all types of electron microscopes. However, there are several advantages of WDS over XEDS:

■ Better energy resolution (5–10 eV) to unravel the peak overlaps that plague XEDS (see Section 34.3).

■ Better peak-to-background capability to detect smaller amounts of elements.

■ Better detection of light elements (minimum $Z = 4$, Be) by careful choice of crystal, rather than solely through a dependence upon electronics as in the XEDS.

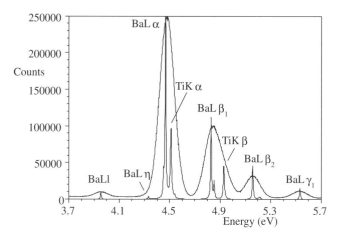

Figure 32.17. A WDS spectrum from BaTiO$_3$, but plotted against energy rather than wavelength. WDS easily resolves the Ba L$_\alpha$/Ti K$_\alpha$ overlap, which is impossible with an EDS as shown in the overlapping spectrum. The improved resolution of WDS (~8 eV) is obvious.

■ Higher throughput count rate using a gas-flow proportional counter.

A typical WDS spectrum from BaTiO$_3$ is shown in Figure 32.17. For comparison, an XEDS spectrum is shown superimposed; the improved resolution and *P/B* for WDS are obvious. Because of these advantages, the WDS is often the spectrometer of choice in the X-ray fluorescence spectrometer (XRF), which has a spatial resolution of a few millimeters, and in the electron probe microanalyzer (EPMA) with a spatial resolution of a few micrometers. The advantages of the WDS may make it an attractive complement to the XEDS in future AEMs. However, WDS systems have not been applied to AEMs because of their low X-ray collection efficiency compared to the XEDS, and we discussed the drawbacks of WDS at the start of the chapter. Only when a WDS is designed that is compact enough to be placed inside an AEM specimen stage will it be possible to realize these advantages, and even then the output count rate from thin specimens might be too low unless the AEM has an FEG (Goldstein *et al.* 1989). Spence and Lund (1991) show preliminary results from a WDS in an AEM giving 40-eV resolution in a study of coherent bremsstrahlung (see Section 33.4.C).

■ No artifacts in the spectrum from the detection and signal processing, except for higher-order lines from fundamental reflections (when $n \geq 2$ in the Bragg equation).

CHAPTER SUMMARY

The XEDS is the only X-ray spectrometer currently used in TEMs. It is remarkably compact, efficient, and sensitive. A combination of Si(Li) and Ge detectors can detect K$_\alpha$ lines from all the elements, from Be to U. However, the XEDS has limits in terms of its need for cooling, its poor energy resolution, and the many artifacts that appear in the spectra. The XEDS is simple to run and maintain if you take care to perform certain basic procedures and refrain from certain others that can damage the detector. Sometimes, it may be too simple; beware. You need to:

■ Measure your detector resolution weekly at the Mn or Cr K$_\alpha$ line (typically, 130–140 eV for Si(Li) and 120–130 eV for IG).
■ Measure the ICC defined by the FWTM/FWHM ratio of the Cr K$_\alpha$ line (ideally 1.82) on a monthly basis.
■ Monitor any ice build-up via the Ni (or Cr) K$_\alpha$/L$_\alpha$ ratio on a weekly basis.
■ Check the calibration of the energy range of your MCA display every few months.
■ Check the dead-time correction circuitry by the linearity of the output count rate versus beam current, every six months.
■ Check the counts in a fixed clock time as a function of beam current, to determine the maximum output count rate, every six months.
■ Be aware of artifacts in *all* your spectra.

For the sake of completeness, Table 32.1 below shows you the relative merits of the various detectors that we have discussed in this chapter.

Table 32.1. Comparison of X-ray Spectrometers

Characteristic energy resolution	Intrinsic Ge	Lithium-drifted Si	WDS
Typical value	140 eV	150 eV	10 eV
Best value	114 eV	127 eV	5 eV
Energy required to form electron–hole pairs (77 K)	2.9 eV	3.8 eV	n.a.
Band-gap energy (indirect)	0.67 eV	1.1 eV	n.a.
Cooling required	LN_2 or thermoelectric	LN_2 or thermoelectric	none
Typical detector active area	10–30 mm^2	10–30 mm^2	n.a.
Typical output counting rates	5–10,000 cps	5–10,000 cps	50,000 cps
Time to collect full spectrum	1 min	1 min	30 min
Collection angle	0.03–0.20 sr	0.03–0.30 sr	10^{-4}–10^{-3} sr
Take-off angle	0°/20°/72°	0°/20°/72°	40°–60°
Artifacts	Escape peaks Sum peaks Ge K/L internal fluorescence peaks	Escape peaks Sum peaks Si K internal fluorescence peaks	High-order lines

REFERENCES

General References

Goldstein, J.I., Newbury, D.E., Echlin, P., Joy, D.C., Romig, A.D. Jr., Lyman, C.E., Fiori, C.E., and Lifshin, E. (1992) *Scanning Electron Microscopy and X-ray Microanalysis,* 2nd edition, Plenum Press, New York.

Heinrich, K.F.J., Newbury, D.E., Myklebust, R.L., and Fiori, C.E., Eds. (1981) *Energy Dispersive X-ray Spectrometry,* NBS Special Publication 604, U.S. Department of Commerce, Washington, DC.

Russ, J.C. (1984) *Fundamentals of Energy Dispersive X-ray Analysis,* Butterworths, Boston, Massachusetts.

Williams, D.B., Goldstein, J.I., and Newbury, D.E., Eds. (1995) *X-Ray Spectrometry in Electron Beam Instruments,* Plenum Press, New York.

Specific References

Goldstein, J.I., Lyman, C.E., and Williams, D.B. (1989) *Ultramicroscopy* **28,** 162.

Joy, D.C. (1995) in *X-Ray Spectrometry in Electron Beam Instruments* (Eds. D.B. Williams, J.I. Goldstein, and D.E. Newbury), p. 53, Plenum Press, New York.

Lund, M.W. (1995) *ibid.,* p. 21.

Lyman, C.E., Newbury, D.E., Goldstein, J.I., Williams, D.B., Romig, A.D. Jr., Armstrong, J.T., Echlin, P.E., Fiori, C.E., Joy, D.C., Lifshin, E., and Peters, K.R. (1990) *Scanning Electron Microscopy, X-Ray Microanalysis and Analytical Electron Microscopy; A Laboratory Workbook,* Plenum Press, New York.

Lyman, C.E., Goldstein, J.I., Williams, D.B., Ackland, D.W., von Harrach, S., Nicholls, A.W., and Statham, P.J. (1994) *J. Microsc.* **176,** 85.

McCarthy, J.J. (1995) in *X-Ray Spectrometry in Electron Beam Instruments* (Eds. D.B. Williams, J.I. Goldstein, and D.E. Newbury) p. 67, Plenum Press, New York.

Michael, J.R. (1995) *ibid.,* p. 83.

Mott, R.B. and Friel, J.J. (1995) *ibid.,* p. 127.

Newbury, D.E. (1995) *ibid.,* p. 167.

Sareen, R.A. (1995) *ibid.,* p. 33.

Spence, J.C.H. and Lund, M. (1991) *Phys. Rev.* **B44,** 7054.

Statham, P.J. (1995) in *X-Ray Spectrometry in Electron Beam Instruments* (Eds. D.B. Williams, J.I. Goldstein, and D.E. Newbury), p. 101, Plenum Press, New York.

Zemyan, S.M. and Williams, D.B. (1994) *J. Microsc.* **174,** 1.

Zemyan, S.M. and Williams, D.B. (1995) *X-Ray Spectrometry in Electron Beam Instruments* (Eds. D.B. Williams, J.I. Goldstein, and D.E. Newbury), p. 203, Plenum Press, New York.

The XEDS–TEM Interface

33

CHAPTER PREVIEW

In principle, all you have to do to create an AEM is to hang an XEDS detector on the side of a TEM. However, in practice it isn't always that simple because the TEM is designed primarily as an imaging tool, and micro-analysis requires different design criteria. The AEM illumination system and specimen stage are rich sources of radiation, not all of it by any means coming from the area of interest in your specimen. So you have to take precautions to ensure that the X-ray spectrum you record comes from the area you chose and can ultimately be converted to quantitative elemental information. You therefore need to understand the problems associated with the XEDS–TEM interface and find ways to maximize the useful data. We describe several tests you should perform to ensure that the XEDS–TEM interface is optimized.

The XEDS–TEM Interface

33

33.1. THE REQUIREMENTS

The interfacing of the XEDS to the TEM is not something over which you have too much control. It has already been carried out by the manufacturers, so you purchase an AEM system of TEM and XEDS. Often you won't be able to change anything about the system, but nevertheless you should be aware of the important factors that characterize the interface between the XEDS and the TEM column and what effect these factors will have on your microanalysis experiments. Knowing these factors may help you to select the best AEM to use.

The stage of a TEM is a harsh environment. An intense beam of high-energy electrons bombards a specimen which interacts with and scatters the electrons. The specimen *and any other part of the microscope* that is hit by electrons emit both characteristic and bremsstrahlung X-rays which have energies up to that of the electron beam. X-rays of such energy can penetrate long distances into material and fluoresce characteristic X-rays from anything that they hit. Ideally, the XEDS should only "see" the X-rays from the beam–specimen interaction volume. However, as shown in Figure 33.1, it is not possible to prevent radiation from the microscope stage and other areas of the specimen from entering the detector. The X-rays from the microscope itself we will call "system X-rays." All the X-rays arising from regions of the specimen other than that chosen for analysis, we term "spurious X-rays." Your job, as an analyst, is to learn how to identify the presence of these undesirable X-rays and to minimize their effect on your microanalysis.

33.2. THE COLLIMATOR

As you can see from Figure 33.1, the XEDS has a collimator in front of the detector crystal. This collimator is the

front line of defense against the entry of undesired radiation from the stage region of the microscope.

The collimator also defines the (desired) collection angle of the detector (see below) and the average take-off angle of X-rays entering the detector.

Ideally, the collimator should be constructed of a high-Z material such as W, Ta, or Pb, coated externally and internally with a low-Z material such as Al, C, or Be. The low-Z coating will minimize the production of X-rays from any backscattered electrons that happen to spiral into the collimator and the high-Z material will absorb any high-energy bremsstrahlung radiation. The inside of the collimator should also have baffles to prevent any backscattered electrons from generating X-rays that then penetrate the detector. Such a design is shown in Figure 33.2 (Nicholson *et al.* 1982), and aspects of this design are available in some commercial systems. They are strongly recommended.

33.2.A. Collection Angle

The detector collection angle (Ω) is the solid angle subtended at the analysis point on the specimen by the active area of the front face of the detector. The collection angle is shown in Figure 33.1 and is defined as

$$\Omega = \frac{A \cos \delta}{S^2} \qquad [33.1]$$

where A is the active area of the detector (usually 30 mm^2), S is the distance from the analysis point to the detector face, and δ is the angle between the normal to the detector face and a line from the detector to the specimen. In many XEDS systems, the detector crystal is tilted toward the specimen so $\delta = 0$; then $\Omega = A/S^2$. It is clear that to maximize Ω the detector should be placed as close to the specimen as possible.

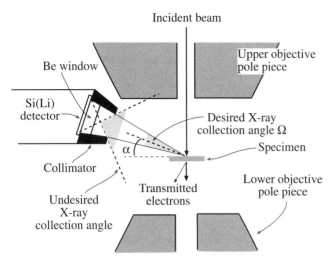

Figure 33.1. The interface between the XEDS and the AEM stage, showing how the detector can "see" undesired X-rays from regions other than the beam–specimen interaction volume. The desired collection angle Ω and take-off angle α are also shown.

> The value of Ω is the most important parameter in determining the quality of your X-ray microanalysis.

In most cases, particularly with thermionic-source instruments, it is the low X-ray counts that limit the accuracy of the experiment. Now in commercial AEMs, S varies from about 10–30 mm, and as a result values of Ω lie in the range from 0.3 down to 0.03 sr. ATW detectors invariably have lower Ω values than Be window or windowless detectors, because the polymer window has to be supported on a grid (usually etched Si) which reduces the

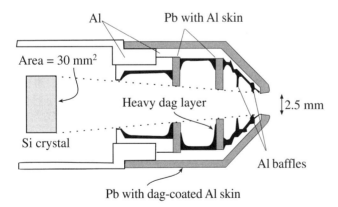

Figure 33.2. Combination high-Z (Pb) and low-Z (Al/carbon paint) collimator design to prevent high-energy bremsstrahlung from penetrating the collimator walls. Baffles are incorporated to minimize BSE entry into the detector.

collection angle by ~20%. So, at best, Ω is a small fraction of the total solid angle of characteristic X-ray generation which is, of course, 4π sr. These values of Ω are calculated from the dimensions of the stage and the collimator. Unfortunately, there is no way you can measure this critical parameter directly, although you can compare X-ray count rates between different detector systems using a standard specimen, such as our thin Cr film, and a known beam current. A figure of merit for this parameter is given in terms of the X-ray counts per second detected from the standard, when a given beam current is used with a given detector collection angle (cps/nA/sr). Typically, for an AEM with a nominal Ω of 0.13 sr and a beam energy of 300 keV the figure of merit is >8000. For an energy of 100 keV, it is about 13,000 (Zemyan and Williams 1994). The increase at lower keV is due to the increased ionization cross section.

The magnitude of Ω is limited because the upper polepiece of the objective lens gets in the way of the collimator, thus limiting S. To avoid this limitation we could increase the polepiece gap, but doing so would lower the maximum beam current and degrade the image resolution, both of which are highly undesirable. So a compromise has to be made in the design of the stage of the AEM to ensure both adequate current in the beam and the best possible collection angle. If we move the detector too close to the specimen, it will eventually suffer direct bombardment by backscattered electrons. The other alternative we have, looking at equation 33.1, is to increase A. However, increasing the detector area results in a small decrease in energy resolution. As we noted already, there are certainly situations where increased count rate is to be preferred at the expense of a small decrease in resolution; while 50-mm² detectors are available, they are rarely used.

33.2.B. Take-Off Angle

The take-off angle α is the angle between the specimen surface (at 0° tilt) and a line to the center of the detector, as shown in Figure 33.1. Sometimes, it is also defined as the angle between the transmitted beam and the line to the detector, which is simply (90° + α). Traditionally in the EPMA, the value of α is kept high to minimize the absorption of X-rays as they travel through bulk specimens. Unfortunately, if we maximize α the price we pay is lowering Ω. Because the detector then has to be positioned above the upper objective polepiece, it will "look" through a hole in the polepiece. Therefore, it will be much further from the specimen. In the EPMA low Ω is not a problem because there are always sufficient X-rays from a bulk specimen, but in the AEM the highest possible Ω is essential, as we've already emphasized.

We would like to optimize the take-off angle and maximize the count rate.

In AEMs where the detector "looks" through the objective polepiece giving a high take-off angle but a low Ω, the poor X-ray count rate makes quantitative analysis much more time-consuming. Keeping the detector below the polepiece restricts α to a maximum value of about 20°. In most cases you will find that such a small value of α is not a problem, because one of the major advantages of thin-specimen AEM compared to bulk EPMA is that absorption can usually be neglected. However, if absorption is a problem in your specimen then you can reduce the path length of X-rays traveling through your specimen by tilting it toward the detector, thus increasing α (see Section 35.5). Tilting may increase spurious effects, which we'll discuss later, and also generally lowers the P/B (peak-to-background) ratio in the spectrum; so tilting is always a compromise.

33.2.C. Orientation of the Detector to the Specimen

(a) Is the detector pointing on axis? The detector is inserted into a position where it is almost touching the objective polepiece, and you hope that it is "looking" at the region of your specimen that is on the optic axis when the specimen is eucentric and at zero tilt. We have to assume that the XEDS and the TEM manufacturers have collaborated closely in the design of the collimator and stage. To find out if your system is well aligned, you can make a low-magnification X-ray map from a homogeneous specimen such as a thin Cr film (Nicholson and Craven 1993). If the detector is not pointing on axis, the map will show an asymmetric intensity distribution. Alternatively, if you cannot map at low enough magnification, simply see how the Cr K_α intensity varies from area to area on the foil with the specimen traverses set at zero and different areas selected using the beam deflectors. The maximum intensity should be recorded in the middle grid square and for some distance around. It is also instructive to do the same test with the specimen moved up or down away from the eucentric plane using the z-control. Again, the maximum intensity should be recorded at the eucentric plane. If the intensity is asymmetric, then the detector or the collimator is not well aligned and some of your precious X-ray intensity is being shadowed from the detector, probably by the collimator; so you need to consult the manufacturer.

(b) Where is the detector with respect to the image? When you look at a TEM or STEM image to position the

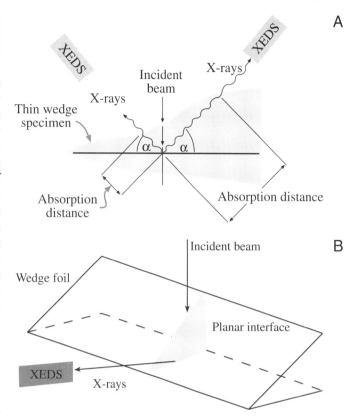

Figure 33.3. (A) The position of the XEDS detector relative to a wedge-shaped thin foil results in different X-ray path lengths. The shortest path length with the detector "looking" at the thinnest region of the foil is best. (B) The preferred orientation of the XEDS detector when analyzing a planar defect: the interface plane is parallel to the detector axis and the incident beam direction.

beam on an area for microanalysis, it is best if the detector is "looking" toward the thin region of the specimen rather than toward the thicker region, as shown in Figure 33.3A. This position minimizes the X-ray path length through the specimen and helps to ensure that any absorption is minimized. In TEM mode, the detector orientation with respect to the BF image on the screen will vary with magnification if the BF image rotates when changing magnification. In a STEM BF image there is no rotation, so the relative orientation of the detector to the image will be fixed. It is simple to find this orientation if the detector axis (y-axis) is normal to a principal traverse axis (x) of the stage. Under these circumstances, if you push in the end of the holder while the specimen is in the column, the image will move in the +x-direction. Then you can determine by simple geometry the direction (+y or −y) along which the detector is looking with respect to that +x-direction in the TEM image. In STEM, the image is sometimes rotated 180° with respect to the TEM image, so you have to take this into account (just check TEM and STEM images of a recognizable area of your specimen).

If you're doing microanalysis across a planar interface, which is a common AEM application, then you will also need to orient your specimen such that the interface is parallel to the detector axis and the beam. A tilt-rotation holder would be ideal for this, but a low-background version is not available, so you may need to reposition your foil manually until the interface is in the right orientation (see Figure 33.3B).

> The XEDS detector must be "looking" at the thin edge of your specimen and aligned with any planar interface you are studying.

33.3. PROTECTING THE DETECTOR FROM INTENSE RADIATION

If you are not careful, the XEDS electronics can be temporarily saturated if high doses of electrons or X-rays hit the detector. The detector itself may also be damaged, particularly in intermediate voltage microscopes. These situations usually occur when you place bulk material under the beam. This can happen if you leave the objective diaphragm inserted, if you go to low magnification and expose the bulk regions of a disk to the beam, or if you are traversing around a thin specimen supported on a grid and a grid bar is hit by the beam. To avoid these problems there are various kinds of shutter systems built into XEDS detectors which automatically protect the detector crystal if the instrument is switched to low magnification or if the pulse processor detects too high a flux of radiation.

> To avoid reliance on the automatic system, it is best to have the shutter closed until you have decided which area you want to analyze, and it is thin enough that the generated X-ray flux doesn't saturate the detector.

If you don't have a shutter, then you can physically retract the detector, which lowers Ω (if it is retracted along a line of sight to the specimen) or removes the detector from out of view of the specimen. The drawback to this approach is that constant retraction and reinsertion of the detector may cause undue wear on the sliding "O"-ring seal and also you may reposition your detector slightly differently each time, unless the system is designed so that you can push the detector up to a fixed stop, thus insuring a constant collection and take-off angle. A shutter is highly recommended!

33.4. SYSTEM X-RAYS AND SPURIOUS X-RAYS

In an ideal AEM, all spectra would be characteristic only of the chosen region of your specimen. The analysis of bulk specimens in the EPMA approaches this ideal, but in the AEM several factors combine to introduce false information which can introduce serious errors into both qualitative and quantitative microanalysis unless you are aware of the dangers and take appropriate precautions to identify and minimize the problems. The factors unique to the AEM that are responsible for these problems are:

- The high accelerating voltages which generate intense doses of stray X-rays and electrons in the illumination system.
- The scattering of high-energy electrons and X-rays by the thin specimen in the limited confines of the TEM stage.

Most AEMs are now designed to minimize some of these problems. Nevertheless, when identification and quantification of small (<~5%) elemental amounts are required, you have to be wary of system and spurious X-rays, which we will now discuss in some detail. These artifacts, which are in addition to the XEDS system artifacts, can be responsible for large errors in quantification or, in the extreme, may make your microanalysis impossible.

33.4.A. Pre-Specimen Effects

Ideally, the electron beam should be the sole source of radiation incident on your specimen, and the X-rays then originate in a well-defined interaction volume. In practice, the illumination system can produce high-energy bremsstrahlung X-rays and uncollimated electrons, which may strike the specimen anywhere, producing spurious X-rays indistinguishable from those generated in the region of interest. In inhomogeneous specimens (which are usually just the kind that we want to analyze) the presence of significant amounts of spurious X-rays means that the quantification process could give the wrong answer. There are several review papers (e.g., Williams and Goldstein 1981, Allard and Blake 1982) which describe in detail how to identify and minimize these artifacts from the illumination system, so we will just describe the precautions necessary to ensure that the AEM is operating acceptably. Since these artifacts are primarily a result of the high-energy electrons interacting with column components such as diaphragms and polepieces, you must take extra care when using intermediate-voltage instruments.

The standard way to detect stray radiation from the illumination system is to position the focused electron beam down a hole in your specimen and see if you can detect an X-ray spectrum characteristic of the specimen.

> Such a spectrum, sometimes termed a "hole count," is *invariably* obtained in all AEMs if you count for long enough.

If the hole count contains more than a few percent of the characteristic intensity obtained from a thin area of your specimen under similar conditions, then we say the illumination system is not "clean."

You can easily determine whether stray electrons or X-rays are the problem, as illustrated in Figure 33.4 (Goldstein and Williams 1978). Almost invariably, the problem is caused by stray X-rays penetrating the C2 diaphragm.

> The solution to this problem is to use very thick (several mm) platinum diaphragms which have a "top-hat" shape and a slightly tapered bore to maintain good electron collimation.

These diaphragms should be a standard fixture in all AEMs (check with your manufacturer), but they are expensive, and you cannot flame-clean them in the usual way. When the thick diaphragms do contaminate, you should discard them, otherwise the contamination itself will become a source of X-rays and also deviate beam electrons by charging. Some AEMs incorporate a small diaphragm just above the upper objective lens to shadow the thicker outer regions of the specimen from stray X-rays. Other AEMs use virtual beam-defining apertures, keeping the diaphragm itself well away from the specimen, and this is ideal. Another good way to minimize the effects of the bremsstrahlung is to use an evaporated film or window-polished flake on a Be grid, rather than a self-supporting disk. If the specimen is thinner than the path length for fluorescence, spurious X-rays will not be generated. Of course, such thin specimens may not be possible to prepare, or may take a great deal of effort, while self-supporting disks are relatively easy and quick to produce; so this isn't a popular suggestion with graduate students.

For a quantitative, reproducible measure of the hole count, you should use a uniform thin specimen such as the Cr film we have described. This film should be supported on a bulk material that has a low-energy (<~3 keV) L line and a high-energy (>~15 keV) K line. A thick molybdenum or gold washer is ideal. Any high-energy bremsstrahlung X-rays penetrating through the C2 diaphragm will strongly fluoresce the Mo K or Au L line, while stray electrons will excite the Mo L or Au M lines preferentially.

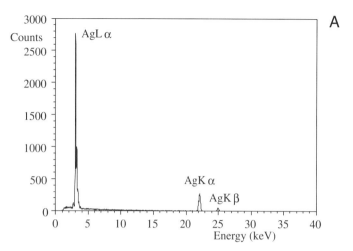

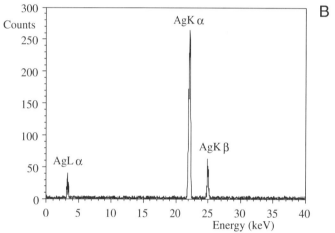

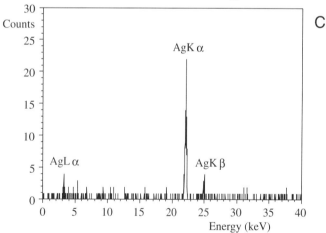

Figure 33.4. The "hole count." (A) A Ag self-supporting disk produces an electron-characteristic (high L/K ratio) spectrum when struck by the primary beam. (B) Without a thick C2 diaphragm, an intense Ag spectrum is also detected when the beam is placed down a hole in the specimen. This spectrum has a low L/K ratio, which indicates high-energy bremsstrahlung fluorescence of the K lines. Approximately 50% of the K_α line in (A) was due to these spurious X-rays. (C) Use of a thick Pt C2 diaphragm reduces the intensity of the hole count substantially. The K_α intensity in (C) is about 30 times less than in (B). (Note the scale change.)

A

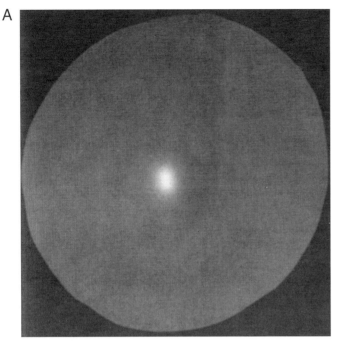

B

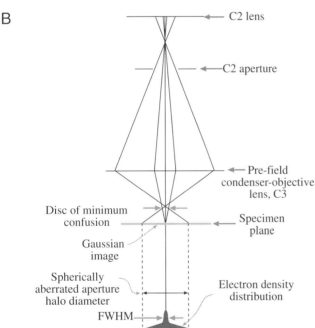

Figure 33.5. (A) The shadow of the diaphragm defines the extent of the halo which excites X-rays remote from the chosen microanalysis region. (B) Ray diagram showing how the STEM probe obtained with a large C2 aperture has a broad halo of electrons surrounding the intense Gaussian central portion. Such a halo is the major source of uncollimated electrons and arises due to spherical aberration in the C3 lens.

As a rule of thumb, the ratio of Mo K_α or Au L_α intensity detected (when the beam is down the hole) to the Cr K_α intensity obtained with the beam on the specimen should be less than 1%.

Under these conditions the remaining stray X-rays will not influence the accuracy of quantification, or introduce detectable peaks from elements not in the analysis region. For more detail on this test see Lyman and Ackland (1991). If you don't want to go to the trouble of this test, then the least you should do is measure the in-hole spectrum from your specimen and subtract it from your experimental spectrum.

In addition to stray X-rays, it is possible that all the electrons are not confined to the beam. If your microscope has a non-beam-defining spray aperture below the C2 aperture, it will eliminate such stray electrons without generating unwanted X-rays. Then the main source of poorly collimated electrons is usually the "tail" of electrons around the non-Gaussian-shaped probe that arises from spherical aberration in the C3 lens, as shown in Figure 33.5 from Cliff and Kenway (1982). The best way to minimize this effect is to image the beam on the TEM screen under the conditions that you will use for microanalysis and select the best C2 aperture, as we discussed earlier in Chapter 6. It is a simple test to move your probe closer and closer to the edge of your specimen and see when you start generating X-rays. Do this with different-size, top-hat C2 diaphragms.

In summary:

■ Always operate with clean, top-hat C2 diaphragms.
■ Use very thin flake specimens, if possible.
■ Always image the electron beam on the TEM screen prior to microanalysis, to ensure that it is well collimated by the C2 aperture.

Under these circumstances, the primary X-ray source will be the region where you put the beam.

33.4.B. Post-Specimen Scatter

After the electrons interact with the specimen, they are scattered elastically or inelastically. It is fortunate for us that the intensity of elastic and inelastic scatter from a thin specimen is greatest in the forward direction. Most of the forward-scattered electrons are gathered by the field of the lower objective polepiece and proceed into the imaging system of the microscope. Unfortunately, some electrons are scattered through high enough angles that they strike some part of the specimen holder or the objective lens polepiece or other material in the stage of the microscope.

> This effect will be severely exacerbated if the objective diaphragm is not removed during microanalysis.

It is instructive to try this experiment (just once!) to see the enormous increase in spurious and system X-rays that result. Usually, the X-ray flux is so great that the pulse-processing electronics are saturated and the dead time reaches 100%, and the automatic shutter will activate. Even when you remove the objective diaphragm, scattered electrons may create X-rays characteristic of the materials used to construct the holder, the polepiece (mainly copper and iron), and the collimator, and these X-rays could be picked up by the XEDS detector. Furthermore, the scattered electrons may travel directly into the XEDS detector, generating electron–hole pairs, or they may hit your specimen at some point remote from the area of interest and produce specimen-characteristic spurious X-rays. These possibilities are undesirable but unavoidable, because without the beam–specimen interactions that produce this scattering, we would get no information at all from the specimen. Figure 33.6 summarizes all the possible sources of spurious X-rays from post-specimen scatter.

In addition to electron scatter, there will be a flux of bremsstrahlung X-rays produced in the specimen. The intensity of these X-rays is also greatest in the forward direction (see shaded area in Figure 33.6). Since they possess a full spectrum of energy, the bremsstrahlung will fluoresce some characteristic X-rays from any material that they strike. The easiest way to discern the magnitude of this problem is to use a uniformly thin foil (such as our standard Cr film) on a copper grid. When you place the probe on the film in the middle of a grid square, many micrometers from any grid bar, the collected spectrum will invariably show a copper peak arising from the grid as a result of interactions with electrons or X-rays scattered by the specimen. An example of this effect is shown in Figure 33.7. You can remove the presence of the copper peak by using a beryllium grid, since Be K_α X-rays are not routinely detectable. However, using Be grids merely removes the observable effect, not the cause. Therefore, the post-specimen scatter will still generate specimen-characteristic X-rays remote from the area of interest, even if a Cu peak is not present.

> Remember that Be oxide is highly toxic if inhaled, so if you have to handle Be grids or other Be components, use gloves and tweezers and don't breathe!

To minimize the effects of the scattered radiation, you should keep your specimen close to zero tilt (i.e., normal to the beam). Experimentally, it seems that if you tilt less than about 10°, then the background intensity is not measurably increased. Under these conditions, your specimen will undergo minimum interaction with both the forward-directed X-rays and any backscattered electrons.

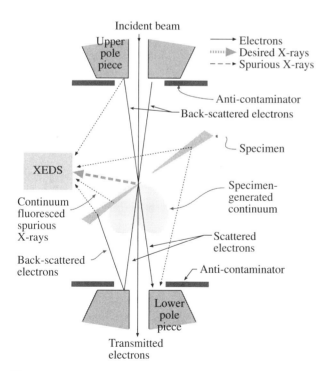

Figure 33.6. Sources of system and spurious X-rays generated when the primary beam is scattered by a tilted, wedge specimen. Note the BSEs which excite X-rays in the stage and elsewhere in the specimen and the specimen-generated bremsstrahlung which fluoresces X-rays from the specimen itself, but well away from the region chosen for analysis.

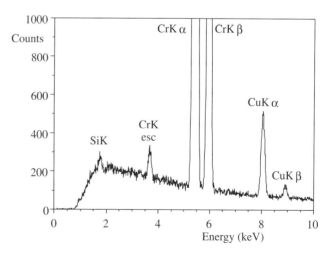

Figure 33.7. Cu peaks in a spectrum from a thin Cr film on a Cu grid. Although the beam is many micrometers from the grid, Cu X-rays are excited by electron scatter and bremsstrahlung from within the specimen and their intensity generally increases with tilt. The Cr escape peak and the Si internal fluorescence peak are also visible.

Both of these phenomena have only a small horizontal component of intensity. The effects of the specimen interacting with X-rays which it has generated will be further reduced if you use thin foils, such as evaporated films or window-polished flakes, rather than self-supporting disks, just as we suggested in the previous section. In self-supporting disks, the bulk regions will interact more strongly with the bremsstrahlung. We do not know what fraction of the post-specimen scatter consists of electrons and what fraction is X-rays, because this will vary with both specimen and microscope conditions. However, there is no evidence to suggest that this X-ray fluorescence limits the accuracy of quantitative analyses.

In addition to keeping your specimen close to zero tilt, you can further reduce the effects of post-specimen scatter by surrounding the specimen with low-atomic-number material. Use of low-Z materials will also remove from your spectrum any characteristic peaks due to the microscope constituents. Be is the best material for this purpose and, as we said right at the beginning of this part of the book, Be specimen holders in addition to Be support grids are essential for X-ray microanalysis. Ideally, all solid surfaces in the microscope stage region that could be struck by scattered radiation should also be shielded with Be. Unfortunately, such modifications are rarely available commercially. The narrow polepiece gap, required to produce high probe currents, and the cold finger, used to reduce hydrocarbon contamination of the specimen, both tend to increase the problems associated with post-specimen scatter. In the ideal AEM, the vacuum would be such that a cold finger would not be necessary and the polepiece gap would be chosen to optimize both the detected peak-to-background ratio and the probe current. When an AEM stage was substantially modified with low-Z material (Nicholson *et al.* 1982), a large reduction in bremsstrahlung intensity was reported and X-ray peak-to-background ratios were produced that are still unmatched by most commercial AEMs. We'll discuss this more in Section 33.5.

You must note, however, that whatever precautions you take, the scattered electrons and X-rays, which are invariably present, result in a specific limitation to X-ray microanalysis.

> If you are seeking small amounts (<~1–2%) of an element A in a specific region of your specimen, and that same element A is present in large amounts, either elsewhere in your specimen or in the microscope stage, then you *cannot* conclusively determine the presence of element A in the specific region of the specimen.

A small peak from A will *invariably* be present in all spectra, just as surely as the Si or Ge internal fluorescence peak will be present.

Obviously, then, you must determine the contributions to the X-ray spectrum from your microscope, and this is best achieved by inserting a low-Z specimen in the beam that generates mainly a bremsstrahlung spectrum, such as an amorphous carbon film, supported on a Be grid or a pure B foil. If a spectrum is accumulated for a substantial fraction of time (say 10–20 minutes), then in addition to the C or B peak, if your XEDS can detect it, the various instrumental contributions to the spectrum should become visible. Such an "instrument spectrum" (see Figure 33.8) should only exhibit the internal fluorescence peak and possibly the Au absorption edge from the detector. Any other peaks will be from the microscope itself, assuming the specimen is pure. These peaks will tell you which elements it is *not* possible to seek in small quantities in your specimen because of their presence in your AEM.

We can summarize the methods used to minimize the effects of post-specimen scattering quite simply:

- Always remove the objective diaphragm.
- Operate as close to zero tilt as possible.
- Use a Be specimen holder and Be grids.
- Use thin foils, flakes, or films rather than self-supporting disks.

Remember that even with these precautions you will still have to look out for artifacts in the spectrum, particularly those from the XEDS system.

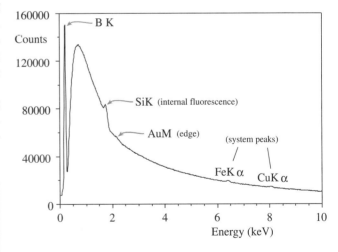

Figure 33.8. An XEDS spectrum from high-purity boron, showing system peaks. The Si K_α peak and the Au M absorption edges are detector artifacts, but the small peaks at 6.4 keV and 8 keV are Fe and Cu system peaks, respectively.

33.4.C. Coherent Bremsstrahlung

As we noted back in Chapter 4, the bremsstrahlung spectrum is sometimes referred to as the "continuum" because the intensity is assumed to be a smooth, slowly varying function of energy. This assumption is perfectly reasonable when the bremsstrahlung is generated in bulk materials by electrons with energies <~30 keV, such as in an SEM. However, in thin monocrystalline specimens illuminated by high-energy electrons, it is possible to generate a bremsstrahlung X-ray spectrum that contains small, Gaussian-shaped peaks known as "coherent bremsstrahlung" (CB). The phenomenon of CB is well known from high-energy physics experiments, but no one thought it would occur at AEM voltages until it was clearly demonstrated by Reese et al. (1984). Figure 33.9A shows a portion of an X-ray spectrum from a thin foil of pure copper taken at 120 keV. The primary peaks, as expected, are the Cu $K_{\alpha/\beta}$ and the L lines. In addition, the es-

cape peak is identified. The other small peaks are the CB peaks. They arise, as shown in Figure 33.9B, by the nature of the coulomb interaction with the regularly spaced nuclei. As the beam electron proceeds through the crystal lattice, close to a row of atoms, each bremsstrahlung-producing event is similar in nature and so the resultant radiation tends to have the same energy. The regular interactions result in X-ray photons of energy E_{CB} given by

$$E_{CB} = \frac{12.4\,\beta}{L\left(1 - \beta\cos\left(90 + \alpha\right)\right)} \qquad [33.2]$$

where β is the electron velocity (v) divided by the velocity of light (c), L is the real lattice spacing in the beam direction (= $1/H$ in a zone-axis orientation), and α is the detector take-off angle. More than one CB peak arises because different Laue zones give different values of L. The CB peak intensity seems greatest when the beam is close to a low-index zone axis, and these conditions should be avoided if possible. Unfortunately, you can't remove the CB effects entirely by operating far from a major zone axis, since some residual peaks are invariably detectable.

> You may mistakenly identify these CB peaks as characteristic peaks from a small amount of some element in the specimen, but fortunately, you can easily distinguish CB peaks from characteristic peaks.

As predicted by equation 33.2, the CB peaks will move depending on both the accelerating voltage (which will alter v and, hence, β) and the specimen orientation, which will change the value of L. Of course, characteristic peaks show no such behavior, and are dependent only on the elements present in the specimen. While the CB peaks are a nuisance, it may be possible to use them to advantage. There is some evidence that the true bremsstrahlung intensity is low in the regions between the CB peaks. Therefore, if you are seeking to detect a very small amount of segregant, e.g., S segregated to grain boundaries in Cu, then it is possible to "tune up" the CB peaks by careful choice of kV and orientation to ensure that the S K_{α} line will appear between two CB peaks and not be masked by them.

33.5. PEAK-TO-BACKGROUND RATIO

The best test of how well your XEDS is interfaced to your TEM is to measure the peak-to-background (P/B) ratio in a standard specimen (the 100-nm Cr film). There are several

A
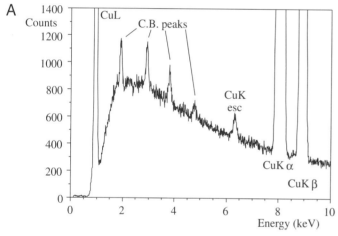

B
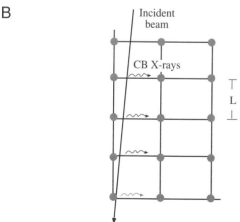

Figure 33.9. (A) CB peaks in a spectrum from pure Cu and (B) the regular generation of CB when the beam passes close to a row of atoms in the specimen.

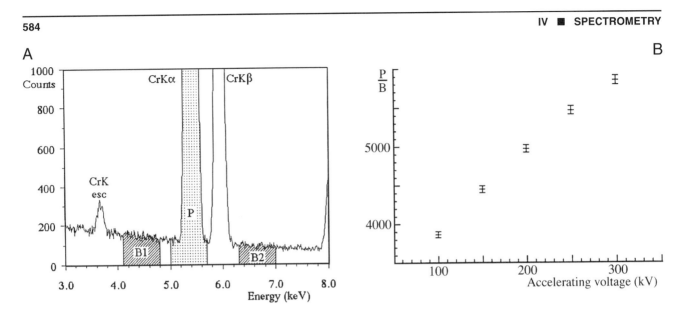

Figure 33.10. (A) The Fiori definition of the peak to background ratio for Cr. The total CrK$_\alpha$ peak intensity P is integrated from 5.0 to 5.7 keV. The background windows B1 and B2 are integrated over seventy 10-eV channels from 4.1 to 4.8 keV and from 6.3 to 7.0 keV, respectively. The average of the two windows [(B1 + B2)/2] is B(avg). This B(avg) is divided by 70 to give the background in a single 10-eV channel [B(10 eV)]. The Fiori definition is given by P/B = [P – B(avg)]/[B(10 eV)]. (B) The increase of the Fiori P/B with accelerating voltage in a well-behaved 300-keV IVEM.

definitions of P/B ratio, but the best one (termed the "Fiori" definition, Fiori *et al.* 1982) is shown in Figure 33.10A. For the Cr K$_\alpha$ peak, you should integrate the peak intensity from 5.0 keV to 5.7 keV and divide this by the average background intensity in a 10-eV channel, as shown in Figure 33.10A. In a well-behaved AEM, the P/B ratio will increase with keV. Recommended P/B values, as shown in Figure 33.10B (Zemyan and Williams 1994), should be close to 4000 at 100 keV, rising to almost 6000 at 300 keV.

CHAPTER SUMMARY

The (S)TEM is not well designed for unambiguous X-ray analysis because X-rays are generated and detected from many sources other than the region of your specimen where you put the beam. Nevertheless, there are well-defined precautions you can take so that you are sure that the spectrum is primarily from your specimen and your interpretation and quantification are not compromised. There are also several standard tests you can carry out to compare your AEM system performance with other instruments.

You must always:

- Ensure the XEDS is pointing toward the thin edge of any wedge specimen.
- Have the shutter closed until you know the area you want to analyze.
- Operate with clean, top-hat C2 diaphragms.
- Use thin foils, flakes, or films rather than self-supporting disks.
- Image the electron beam on the TEM screen to ensure that it is Gaussian.
- Remove the objective diaphragm.
- Operate as close to zero tilt as possible.

Check that:

- The hole count is <1%.
- You know your system peaks and other artifacts.

REFERENCES

General References

Goldstein, J.I. (1979) in *Introduction to Analytical Electron Microscopy* (Eds. J.J. Hren, J.I. Goldstein, and D.C. Joy), p. 83, Plenum Press, New York.

Goldstein, J.I., Williams, D.B., and Fiori, C.E. (1986) in *Principles of Analytical Electron Microscopy* (Eds. D.C. Joy, A.D. Romig Jr., and J.I. Goldstein), p. 123, Plenum Press, New York.

Williams, D.B. (1987) *Practical Analytical Electron Microscopy in Materials Science,* 2nd Edition. Philips Electron Optics Publishing Group, Mahwah, New Jersey.

Zaluzec, N.J. (1979) in *Introduction to Analytical Electron Microscopy* (Eds. J.J. Hren J.I. Goldstein, and D.C. Joy), p. 121, Plenum Press, New York.

Specific References

Allard, L.F. and Blake, D.F. (1982) in *Microbeam Analysis-1982* (Ed. K.F.J. Heinrich), p. 8, San Francisco Press, San Francisco.

Cliff, G. and Kenway, P.B., *ibid.,* p, 107.

Fiori, C.E., Swyt, C.R., and Ellis, J.R., *ibid.,* p. 57.

Goldstein, J.I. and Williams, D.B. (1978) in *SEM-1978* **1**, (Ed. O. Johari), p. 427, SEM Inc., AMF O'Hare, Illinois.

Lyman, C.E., and Ackland, D.W. (1991) in *Proceedings of 49th EMSA Meeting* (Ed. G.W. Bailey and E.L. Hall), p. 720, San Francisco Press, San Francisco.

Nicholson, W.A.P. and Craven, A.J. (1993) *J. Microsc.* **168**, 289.

Nicholson, W.A.P., Gray, C.C., Chapman, J.N., and Robertson, B.W. (1982) *J. Microsc.* **125**, 25.

Reese, G.M., Spence, J.C.H., and Yamamoto, N. (1984) *Phil. Mag.* **A49**, 697.

Williams, D.B. and Goldstein, J.I. (1981) in *Energy Dispersive X-ray Spectrometry* (Eds. K.F.J. Heinrich, D.E. Newbury, R.L. Myklebust, and C.E. Fiori), p. 341, NBS Special Publication 604, U.S. Department of Commerce/NBS, Washington, DC.

Zemyan, S.M. and Williams, D.B. (1994) *J. Microsc.* **174**, 1.

Qualitative X-ray Analysis

34

CHAPTER PREVIEW

It is a waste of time to proceed with *quantitative* microanalysis from your XEDS spectrum without first carrying out *qualitative* analysis. First we will show you how to choose the best operating conditions, for both the microscope and the XEDS system. Then we will explain the best way to obtain a spectrum for qualitative analysis. Qualitative analysis requires that every peak in your spectrum be identified unambiguously, and with statistical certainty, otherwise it should be ignored. So you have to acquire a spectrum with sufficient X-ray counts to allow you to draw the right conclusions with confidence. There are a couple of simple rules to follow which allow you to do this.

> Although such an approach may seem time-consuming and unnecessarily tedious, the need for initial *qualitative* analysis of the spectrum cannot be stressed too strongly.

Two advantages are gained from this approach. First, you may be able to solve the problem at hand without the necessity of a full quantification routine. Second, when quantification is carried out, you will not spend an inordinate amount of time analyzing an artifact or a statistically insignificant peak, and you can be confident that your results are valid.

Qualitative X-ray Analysis

34

34.1. MICROSCOPE VARIABLES

When you first acquire a spectrum, the operating conditions should maximize the X-ray count rate to give you the most intensity in the spectrum, in the shortest time, with the minimum of artifacts. In addition, you want to get a good idea of which elements in your specimen are detectable. The best conditions for qualitative analysis require that you obtain the spectrum from a large area of the specimen, using a large probe, and so spatial resolution will be poor. Having carried out qualitative analysis of a relatively large region, you may then wish to do further analysis of smaller areas, under conditions that optimize spatial resolution, which we discuss in Chapter 36.

To get the most X-ray counts in your spectrum, use the highest operating voltage, since this gives the highest brightness. Notice that in Chapter 33 we said that you get a higher count rate if you decrease the kV because the scattering cross section, σ, increases when the kV decreases; that was a specimen effect. Now we are talking about a gun effect; as the kV increases, the gun brightness increases. While the two effects counter each other, the added advantage of increased peak-to-background ratio with increased kV tips the scales in favor of using the maximum kV. Only choose a lower voltage if knock-on damage is a problem, as might be the case, for example, in a 200-kV to 400-kV instrument. Pick a portion of your specimen that is single phase in the area of interest and is well away from strong diffraction conditions, so as to minimize crystallographic effects and coherent bremsstrahlung. You will need a probe current of several tens of nanoamps. The necessary combination of probe size and final aperture obviously depends on the type of source in your microscope. To get several tens of nanoamps, from a thermionic source, you will have to select a relatively large probe size, say several tens of nanometers, and a large C2 aperture. As we shall see later, these are just the opposite of the requirements for mi-

croanalysis at high spatial resolution. In contrast to the limitations of a thermionic source, under most operating conditions an FEG source will give sufficient current for both initial qualitative analysis and subsequent quantitative analysis for high spatial resolution. However, the lower current from an FEG may mean that you have to count X-rays for a longer time compared to a thermionic source.

You can always gather a more intense spectrum by choosing a thicker region of the specimen. There is nothing wrong with doing this when you are carrying out *qualitative* microanalysis. The only danger is that, if you have a few weight percent of a light element present in the sample, the X-rays may be absorbed within the specimen and so may not be detected. Also, a thick specimen degrades the spatial resolution, but we've already agreed to compromise that aspect of the microanalysis during the initial qualitative analysis.

So, good qualitative analysis requires a large number of X-ray counts in the spectrum. These counts take a long time to generate, so you run the danger of damaging or changing the chemistry of beam sensitive specimens. You may also contaminate the chosen area. To minimize these effects you should spread the beam over as large an area as possible, either by overfocusing C2 if you're in TEM mode or by rastering the beam in STEM mode. Use a liquid-N_2-cooled low-background holder if contamination is still a problem.

34.2. ACQUIRING A SPECTRUM

The first and most important step in qualitative analysis is to acquire a spectrum across the complete X-ray energy range. Microanalysis can often be accomplished using X-rays with energies from ~1–10 keV, and this is the typical range used in the SEM. However, the TEM has a much higher accelerating voltage, and the consequent increase in available overvoltage means that you can easily generate and detect much higher energy X-rays. If you are using an intermediate voltage AEM and a windowless IG detector, we noted in Chapter 32 that all the possible K_α lines from all the elements above Be in the periodic table can be detected.

> The first thing to do is to adjust your MCA system to display the widest possible energy range. For a Si(Li) detector, 0–40 keV is sufficient, and for an IG detector, 0–80 keV may be useful.

Of course, if you know the specimen you are analyzing, such a step may not seem essential, but it is still a wise precaution since unanticipated contaminants or trace impurities may be present. Collect a spectrum for several hundred seconds and ascertain the actual energy range over which all the characteristic peaks occur. Then, if all the peaks are present in an energy range <40 keV, regather the spectrum over that reduced range, thereby improving the resolution of your MCA display by lowering the number of eV per channel. The spectrum that you finally gather for qualitative analysis should be displayed with *at least 10 eV per channel* resolution on the MCA, and a display range of 0–20 keV should be possible under these conditions (i.e., 2048 channels in total).

You can also increase the intensity of the spectrum by lowering the detector time constant to maximize the throughput of counts. This step also degrades the energy resolution of the XEDS but, for many qualitative analyses, this is not important.

> You should use the shortest time constant while maintaining adequate resolution to discriminate the characteristic peaks in the spectrum.

Watch the dead-time readout while acquiring the spectrum to make sure you haven't chosen a combination of probe current and specimen thickness that overloads the detector electronics. Remember that you want to keep the dead time below about 50–60%, and an output count rate of around 5000–10,000 cps is about the best that can be handled by most analog detector electronics under these conditions.

Remember that we have been talking about several different "resolutions." Don't confuse them.

■ Spatial resolution   distances measured in nm (see Chapter 36).
■ Chemical resolution   detectability depending on P/B (see Chapter 36).

■ Energy resolution   identifying elements by distinguishing peaks; different eV (this chapter).

When you've got a good intense spectrum over a suitable energy range, there is a well-defined sequence of steps that should be followed to ensure that you correctly identify each peak in the spectrum and disregard those peaks that are not statistically significant. The computer system can be used to run an automatic identification check on the peaks in the spectrum, assuming the energy display is well calibrated. If the spectrum is simple, containing a few well-separated peaks, this automatic step may be all that is required. However, if your spectrum contains many peaks, and particularly if peak overlap is occurring, then misidentification may occur during such an "autosearch" routine, especially if there is no operator intervention. In addition, small peaks may sometimes be missed and phenomena such as coherent bremsstrahlung may not be taken into account. Under these circumstances there is a well-established manual sequence, developed for analysis of spectra from the SEM by Goldstein *et al.* (1992), and we will describe a modified form of this procedure in the next section.

34.3. PEAK IDENTIFICATION

First of all, we assume that you know what system peaks, if any, occur in your AEM, and what artifacts are likely to occur in your XEDS system. Now, ensure that the MCA display is calibrated to be accurate at the display resolution over the energy range you selected. So if your spectrum is displayed at 10 eV per channel, the peak centroids must be within 10 eV of their true position on the energy scale.

> The key to good qualitative analysis is to be suspicious and to not just seek the peaks you expect, but to be prepared also to find peaks that you don't expect.

Our peak analysis will always include three steps:

■ Look at the most intense peak and work on down through its family; this is just bookkeeping.
■ Go to the next most intense peak not included in the previous step and repeat the search. Then repeat this exercise until all peaks are identified.
■ Think about pathological overlaps; look for spurious peaks, system peaks, and artifact peaks.

The bookkeeping. Starting from the high-energy end of the spectrum, select the most intense peak and determine the possible K, L, or M lines that could be present at that energy, either by using the computer-generated X-ray line markers on the MCA screen or by consulting an appropriate source, such as the "slide rules" offered by most commercial manufacturers. Good "bookkeeping" is essential during the sequence we will now describe, particularly if the spectrum contains many peaks. You must take care to label each peak when you have decided on its source. Proceed as follows:

■ If a K_α line matches the peak, look for the K_β line which has ~10–15% of the K_α intensity. The K_β line *must* be present at X-ray energies above ~2.3 keV (S K_α), but below this energy the detector may not be able to resolve the two lines.

■ If a K_α and K_β pair fit the peaks and the K_α energy is >~8 keV (Ni K_α), look for the L lines at ~0.9 keV if you are using a Be window detector. For a UTW detector, the L_α lines from Cl and above (>0.2 keV) may be detectable. Ni L_α = 849 eV, Cl L_α = 200 eV.

■ If a K_α line does not fit, check for an L_α or M_α line fit.

■ If an L_α line fits, there *must* be accompanying lines in the L family. The number of visible lines will vary, depending on the intensity and energy of the L_α line. The other lines in the family are all of lower intensity than the L_α line, and the following lines may be detectable (the number in parentheses is the intensity relative to the L_α line): $L_{\beta1}$ (0.7), $L_{\beta2}$ (0.2), and $L_{\gamma1}$ (0.08) lines at higher energies and possibly the L_ℓ (0.04) line at lower energy. Other, even less intense lines ($L_{\gamma3}$ (0.03) and L_η (0.01)) may be visible if the L-line family is extraordinarily intense, but this is rare.

■ If the L lines fit, there *must* be a higher-energy K_α/K_β pair, assuming the beam energy is sufficient to generate the K lines and the MCA energy range is wide enough to display them.

■ The M lines are only usually visible for elements above La in the periodic table if a Be window detector is used, and above about Nb if a UTW detector is used. La M_α = 833 eV, Nb M_α = 200 eV.

■ The M_α/M_β line overlap is difficult to resolve because all the M lines are below 4 keV. If an M_α/M_β line fits, look for three very small M_ζ (0.06), M_γ (0.05), and $M_{II}N_{IV}$ (0.01) lines .

■ If the M_α line fits, there *must* be a higher-energy L-line family and possibly the very high energy K lines may exist; again, this depends on the detector, MCA display, and the accelerating voltage.

Figure 34.1 shows the families of lines expected in the display range from 0 to 10 keV, giving you some idea of the distribution of families of elemental lines that you should expect when you follow the procedure outlined above. For example, you can see for which elements you should expect to see a single K line or resolve the K_α/K_β pair, and for which elements you should expect to see both K and L families or L and M families.

> The idea is that you are looking for families of peaks. If a family member is missing your identification may be wrong.

Repeat the exercise. Go to the next most intense peak that has not been identified by the eight steps in the first search. Continue this process *until all the major peaks are accounted for.* Finally, look for the escape peak(s) and sum peak associated with all *major* characteristic peaks that you have conclusively identified. Remember that these artifacts and any CB peaks will be very small, and before you worry about them you should make sure that the peaks are statistically significant; we discuss how to do this for all minor peaks in Section 34.4 below. If you have a Si(Li) detector, the Si escape peaks will lie at 1.74 keV below major peaks in the spectrum and will not occur for elements below phosphorus. For an IG detector, there may be both Ge K- and L-line escape peaks at the appropriate energy below major peaks (9.89 keV for the Ge K_α escape and (much less intense) 1.19 keV for the Ge L_α escape). If you suspect a sum peak at twice the energy of any major peaks, then reacquire the spectrum at a much lower dead time (<20%) and see if the suspected sum peak disappears. If you suspect a CB peak, then reacquire the spectrum at a different accelerating voltage or specimen orientation and see if the small peak shifts.

Check for special cases. The relatively poor energy resolution of the XEDS detector means that there are several pairs of peaks that occur quite commonly in materials science samples that cannot be resolved. These are called "pathological overlaps" and include (a) the K_β and K_α lines of neighboring transition metals, particularly Ti/V, V/Cr, Mn/Fe, and Fe/Co; (b) the Ba L_α line (4.47 keV) and the Ti K_α line (4.51 keV); (c) the Pb M_α (2.35 keV), Mo L_α (2.29 keV), and S K_α (2.31 keV) lines; and finally (d) the Ti, V, and Cr L_α lines (0.45–0.57 keV) and the K lines of N (0.39 keV) and O (0.52 keV) detected in UTW systems.

> *Pathological overlap:* When it is impossible to separate two peaks even when you know they are both there.

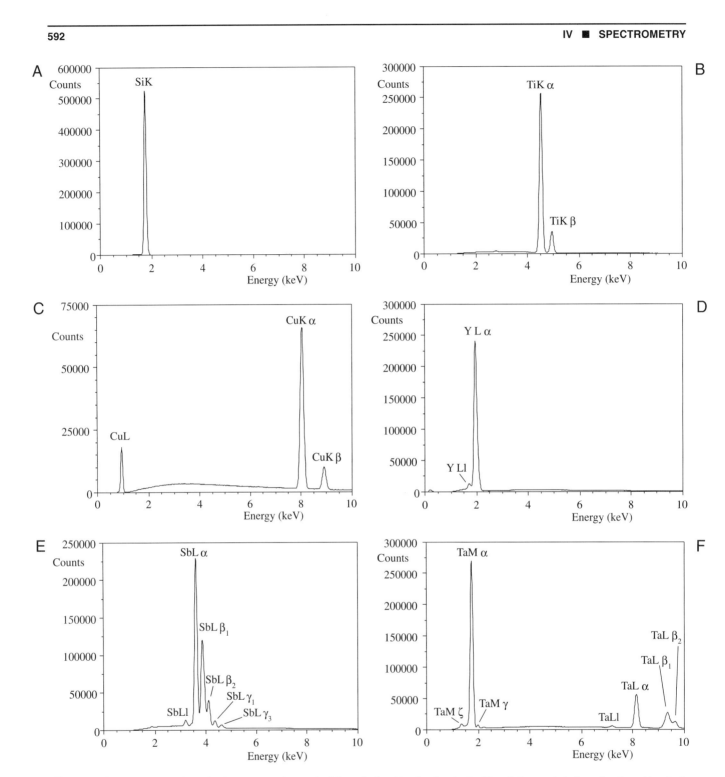

Figure 34.1. X-ray spectra from six elements spanning much of the periodic table, showing the families of characteristic lines. Starting with a single K_α line at low Z and low X-ray energy, the series progresses through the appearance of the L- and M-line families. Note the increasing separation of the lines in a given family as Z and keV increase.

These problems can sometimes be approached by careful choice of the MCA display range. For example, if you are only observing from 0–10 keV, the S K/Mo L line overlap would be clarified by the presence or absence of the Mo K lines around 18 keV. If you suspect that these or any other pathological peak overlap is occurring in your spectrum, then the first thing to do is to regather the spectrum under conditions that maximize the energy resolution of the detector system (i.e., longest time constant and count rate below 5000 cps), and also maximize the display resolution to at least 5 eV per channel. If the overlap is still not resolvable, then you should run a peak deconvolution routine that should be present in the available computer programs (e.g., Schamber 1981). Such routines are capable of detecting and resolving many of the classical materials science overlaps such as Mo L_α and S K_α, and a schematic deconvolution is shown in Figure 34.2.

A historical aside: peak deconvolution has been of great concern to microanalysts since the earliest days of EPMA, using pulse-height analyzers, and the first primitive attempt to deconvolute overlapping peaks was developed by Dolby (1959). With hindsight, it is not difficult to make the connection between the problems of extracting peak information from a low-resolution spectrum and the problem of extracting clear sound signals from noisy recording tape. Dolby, however, saw the potential before anyone else and went on to commercialize his ideas with resounding success, prompting his Ph.D. supervisor Ellis Cosslett to remark that Dolby was the only graduate student he had known to become a millionaire from his Ph.D. research!

This procedure should permit you to identify all the major peaks in the spectrum, but there may still be minor peaks which may or may not be significant and you have to decide whether you are going to identify or ignore these peaks. We'll tell you how to make this decision in the next section.

34.4. PEAK VISIBILITY

Small-intensity fluctuations that you cannot clearly identify as peaks are often present in your spectrum. In this case, there is a simple statistical criterion (Liebhafsky *et al.* 1972) that you can apply to ascertain if the peak is statistically significant or if it can be dismissed as random noise. You must count for a long enough time so that the bremsstrahlung intensity is relatively smooth and any peaks are clearly visible, as summarized in Figure 34.3.

- ■ Increase the display gain until the average background intensity is half the total full scale of the display, so the small peaks are more easily observed.
- ■ Get the computer to draw a line under the peak to separate the peak and background counts.
- ■ Integrate the peak (I_A) and background (I_A^b) counts over the same number of channels; use FWHM if it can be discerned with any confidence; if not, then the whole peak integral will do.

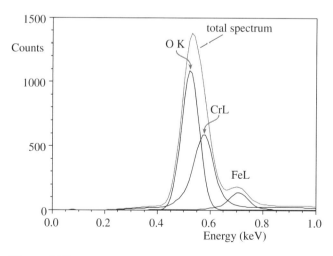

Figure 34.2. The total spectrum arises from the overlap of three Gaussian spectral peaks (the L_α lines of Fe and Cr and the O K_α line) from a mixed Fe-Cr oxide. Deconvolution is essential to determine the intensities in the three constitutent peaks, prior to any attempt at quantification.

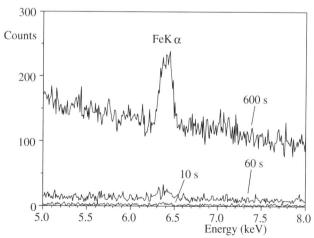

Figure 34.3. With increasing counting time a clear characteristic Fe K_α peak develops above background in a spectrum from Si-0.2% Fe. This demonstrates the need to acquire statistically significant counts before deciding if a small peak is present or absent.

> If $I_A > 3\sqrt{I_A^b}$, then the peak is statistically significant at the 99% confidence limit and must be identified. You will make an erroneous peak identification in less than 1% of analyses using this criterion.
> If $I_A < 3\sqrt{I_A^b}$, then the peak is not significant and should be ignored.

If the insignificant peak is at an energy where you expect a peak to be present, but you think there is only a small amount of the suspected element in your sample, then *count for a longer time* to see if the statistical criterion can be satisfied in a reasonable length of time. If this peak is a critical one, and it is often the minor or trace elements that are most important, then take whatever time is necessary to detect the peak.

> There is no reason not to gather the spectrum for many minutes or even an hour or more, as long as doing so does not change or contaminate your specimen.

However, do *not* obtain more counts by raising the count rate above that which the processing electronics can handle, because you may introduce extra sum peaks and also degrade the energy resolution of the spectrum. Be aware that when you count for long periods of time to search for characteristic peaks of low intensity, you will also begin to detect the small peaks that arise from the various spurious effects that were discussed in detail above, e.g., CB peaks, Si or Ge internal-fluorescence peaks, and system peaks such as Fe and Cu. Also, you increase the possibility of contamination and beam damage to your specimen. So, as we stated at the beginning, it is best to spread the probe over as large an area as possible either by defocusing the C2 lens in TEM mode or by using a scan raster in STEM mode.

Identifying the statistically significant peaks by the above method is one thing. Quantifying the amount of the element responsible for the peak is another matter, and usually many more counts are required, as we'll see when we talk about detectability limits in Chapter 36. However, you may be able to identify the phase that is being analyzed without any further work. For example, in the material that you are investigating, there may be only a few possible phases that can exist after the processing/thermal treatment given to it, and these phases may have very different chemistries. A glance at the relative peak intensities may be sufficient to conclude which phase you have just analyzed because, as we shall see in the next chapter, one of the marvelous advantages of thin-foil microanalysis in the AEM is that the peak intensities are often directly pro-

portional to the elemental concentrations. As a result, quantification can be extremely simple.

To conclude this section, we'll look at two examples:

An example of qualitative microanalysis is shown in the spectrum in Figure 34.4. The spectrum is from a thin NIST oxide glass film on a carbon support film on a Cu grid. X-rays were accumulated for 1000 s with a Be window IG detector at an accelerating voltage of 300 kV. Because of the Be window, we do not expect to see lines below ~0.8 keV and so the O K_α (0.52 keV) will not be detectable. The spectrum only contained peaks in the range from 0–10 keV and the first peak to be examined was the most intense high-energy peak, labeled #1 at an energy of 6.4 keV. The K-line markers identified it, along with the smaller one to its right, as being the Fe K_α and K_β pair. No L-line fit was reasonable (Dy L_α at 6.5 keV being the only alternative). A similar treatment of the next most intense high-energy line (#2) at 3.69 keV produced a match with the Ca K_α (and K_β) and line #3 was consistent with the Si K_α line at 1.74 keV (the K_α/K_β pair cannot be resolved at this energy). Next, the smaller peaks were tackled and the Cu K_α (and K_β) was identified at 8.04 keV, the Ar K_α at 2.96 keV (the K_β was too small to be visible), and the Mg K_α was the last to be identified at 1.25 keV. No escape or sum peaks were detectable, but since the specimen was on a Cu support grid the Cu peaks are most probably due to post-specimen scatter of electrons or X-rays, and we cannot conclude that there is any Cu within the specimen.

> The absence of the Cu L_α line at 0.93 keV is evidence that the thick Cu grid is responsible, since the low-energy L X-rays will be absorbed in the grid itself before they can be detected.

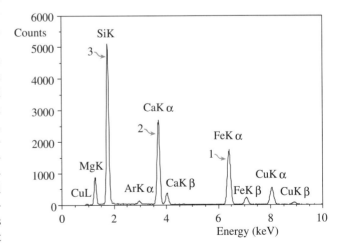

Figure 34.4. Energy-dispersive spectrum obtained at 300 kV from a thin oxide glass film on a Cu grid, with the characteristic peaks identified through the procedure outlined in the text.

Another example is shown in Figure 34.5 and this spectrum contains six Gaussian peaks, which can easily be identified following the procedure outlined above as the K_α and K_β pairs from Fe, Cr, and Ni. To the average metallurgist, this sample can only be some kind of stainless steel and this may be all the information that is required, making subsequent quantitative analysis redundant. But if more information is required, such as the specific grade of stainless steel, then it is necessary to make some measurements of the relative peak intensities, and this is the first step in the quantification procedure. In fact, we will see that the quantification equation, to a first approximation, predicts that the amount of each element is directly proportional to the peak height, and so measuring the relative heights of the K_α peaks in Figure 34.5 with a ruler will give an estimate of the composition as ~Fe-20% Cr-10% Ni. A full quantification using the procedures described in the next chapter gives a very similar result, but with much greater confidence in the true composition.

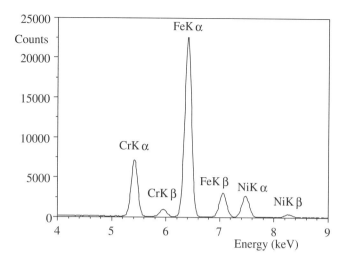

Figure 34.5. Spectrum from a stainless-steel foil. From such a spectrum, in which the peaks are resolved and close together in energy, a first-approximation quantification is possible simply by measuring the relative heights of the K_α peaks.

CHAPTER SUMMARY

One last time: doing the qualitative analysis first is not an option. It is essential.

- Get an intense spectrum across the energy range that contains all the characteristic peaks.
- Starting at the high-energy end of the spectrum, identify all the major peaks and any associated family lines and artifacts.
- If in doubt, collect for a longer time to decide if the intensity fluctuations are in fact peaks.
- Beware of pathological overlaps and be prepared to deconvolute any that occur.

REFERENCES

General References

Goldstein, J.I., Newbury, D.E., Echlin, P., Joy, D.C., Romig, A.D. Jr., Lyman, C.E., Fiori, C.E., and Lifshin, E. (1992) *Scanning Electron Microscopy and X-ray Microanalysis,* 2nd edition, p. 341, Plenum Press, New York.

Russ, J.C. (1984) *Fundamentals of Energy Dispersive X-Ray Analysis,* Butterworths, Boston.

Specific References

Dolby, R.M. (1959) *Proc. Phys. Soc. London* **73**, 81.

Liebhafsky, H.A., Pfeiffer, H.G., Winslow, E.H., and Zemany, P.D. (1972) *X-rays, Electrons and Analytical Chemistry,* p. 349, John Wiley and Sons, New York.

Schamber, F.H. (1981) in *Energy Dispersive X-ray Spectrometry* (Eds. K.F.J. Heinrich, D.E. Newbury, R.L. Myklebust, and C.E. Fiori), p. 193, NBS Special Publication 604, U.S. Department of Commerce/NBS, Washington, DC.

Quantitative X-ray Microanalysis

35

CHAPTER PREVIEW

You've now got an idea of how to acquire XEDS data from thin foils. You understand what factors may limit the information in them and what false and misleading effects may arise. Also, you know how to be sure that a certain peak is due to the presence of a certain element and the occasions when you may not be so sure. Having obtained a spectrum that is qualitatively interpretable, it turns out to be a remarkably simple procedure to convert that spectrum into quantitative data about the elements in your specimen, and this is what we describe in this chapter.

    This chapter is rather long. You will find that you can skip parts of it as you work through it the first time. We have decided to keep the material together so as to be a more useful reference when you are actually doing the analysis on the microscope.

Quantitative X-ray Microanalysis

35

35.1. HISTORICAL PERSPECTIVE

Quantitative X-ray analysis in the AEM is a most straightforward technique. What is surprising is that, given its simplicity, relatively few users take the trouble to extract quantitative data from their spectra, despite the fact that numerical data are the basis for all scientific investigations. Before we describe the steps for quantification, you should know a little about the historical development of quantitative X-ray microanalysis, because this will emphasize the advantages of thin-foil microanalysis over analysis of bulk specimens.

Historically, X-ray microanalysis in electron-beam instruments started with the study of bulk specimens in which the electron beam is totally absorbed, as opposed to "thin" specimens through which the beam penetrates. The possibility of using X-rays generated by a focused electron beam to give elemental information about the specimen was first described by Hillier and Baker (1944), and the necessary instrumentation was built several years later by Castaing (1951). In his extraordinary Ph.D. dissertation, Castaing not only described the equipment but also outlined the essential steps to obtain quantitative data from bulk specimens. The procedures that Castaing proposed still form the basis of the quantification routines used today in the EPMA and may be summarized as follows. Castaing assumed that the concentration C_i of an element i in the specimen generates a certain intensity of characteristic X-rays. However, it is very difficult in practice to measure this generated intensity so Castaing suggested that a known standard of composition $C_{(i)}$ be chosen for element i. We then measure the intensity ratio $I_i/I_{(i)}$, where I_i is the measured intensity emerging from (*not* generated within) the *specimen* and $I_{(i)}$ is the measured intensity emerging from the *standard*.

Castaing then proposed that, to a reasonable approximation

$$C_i / C_{(i)} = [K] I_i / I_{(i)} \qquad [35.1]$$

where K is a sensitivity factor that takes into account the difference between the generated and measured X-ray intensities for both the standard and the unknown specimen. The contributions to K come from three effects:

- Z The atomic number.
- A The absorption of X-rays within the specimen.
- F The fluorescence of X-rays within the specimen.

The correction procedure in bulk microanalysis is often referred to as the ZAF correction. The necessary calculations, which have been refined over the years since Castaing first outlined them, are exceedingly complex and best handled by a computer. (If you're interested, there are several standard textbooks available which describe the ZAF and related procedures in detail, for example, Heinrich and Newbury 1991.)

It was soon realized that if a thin electron-transparent specimen was used rather than a bulk specimen, then the correction procedure could be greatly simplified because, to a first approximation, the A and F factors could be ignored and only the Z correction would be necessary. In addition, if thin specimens were used, the analyzed volume would be substantially reduced, giving a much better spatial resolution. (We discuss this latter point in detail in the next chapter.)

These two obvious advantages of thin-foil microanalysis led to the development of the so-called electron microscope microanalyzer (EMMA), pioneered by Duncumb in England in the 1960s. Unfortunately the EMMA was ahead of its time, mainly because the WDS was the only X-ray detector system available. As we have seen, the WDS is handicapped by its poor collection efficiency, relatively cumbersome size, and slow, serial operation. These factors, particularly the poor efficiency, meant that a large probe size (~0.2 μm) had to be used to generate sufficient X-ray intensity for quantification, and therefore the gain in spatial resolution over the EPMA was not so great. Also, the poor stability of the WDS meant that it was necessary to measure the beam current to make sure that the X-ray intensities from both standard and unknown could be sensibly compared. As a result of all these drawbacks, the EMMA never sold well and the manufacturer (AEI) went out of the EM business.

It is ironic that around this time the commercial developments that would transform TEMs into viable AEMs were all taking place. We've seen that the XEDS detector was developed in the late 1960s, and about the same time the development of commercial TEM/STEM systems was beginning. However, before the demise of the EMMAs, they were to play a critical role in the development of the thin-foil microanalysis procedures that we use today. The EMMA at the University of Manchester, operated by Cliff and Lorimer, was refitted with an XEDS system and they soon realized that the pseudo-parallel collection mode, the greater collection efficiency, and the improved stability of the XEDS removed many of the problems associated with WDS on the EMMA. Cliff and Lorimer (1975) showed that quantification was possible using a simplification of Castaing's original ratio equation, in which there was no need to incorporate intensity data from a standard, but simply ratio the intensities gathered from two elements simultaneously in the XEDS. This finding revolutionized thin-foil microanalysis.

35.2. THE CLIFF–LORIMER RATIO TECHNIQUE

The basis for this technique is to rewrite equation 35.1 for two elements A and B in a binary system.

- We have to measure the above-background characteristic intensities, I_A and I_B, simultaneously. This is trivial with an XEDS and therefore there is no need to measure the intensity from a standard.

- We assume that the specimen is thin enough so that we can ignore any absorption or fluorescence. This assumption is called the *thin-foil criterion*.

The weight percents of each element C_A and C_B can be related to the measured intensities by the so-called Cliff–Lorimer equation

$$\frac{C_A}{C_B} = k_{AB}\frac{I_A}{I_B} \qquad [35.2]$$

The term k_{AB} is often termed the Cliff–Lorimer factor. It is actually *not* a constant, so don't be fooled by the use of "*k*." It varies according to the TEM/XEDS system and the kV, as we will see later. Because we are ignoring the effects of absorption and fluorescence, k_{AB} is related to the atomic-number correction factor (Z) in Castaing's original ratio equation. (We will derive this relationship rigorously in Section 35.3.B.) Now to obtain an absolute value for C_A and C_B we need a second equation, and in a binary system we simply assume that A and B constitute 100% of the specimen, so

$$C_A + C_B = 100\% \qquad [35.3]$$

We can easily extend these equations to ternary and higher-order systems by writing extra equations of the form

$$\frac{C_B}{C_C} = k_{BC}\frac{I_B}{I_C} \qquad [35.4]$$

and

$$C_A + C_B + C_C = 1 \qquad [35.5]$$

You should also note that the *k* factors for different pairs of elements AB, BC, etc., are related, thus

$$k_{AB} = \frac{k_{AC}}{k_{BC}} \qquad [35.6]$$

> It is a convention that we define the units of composition as wt.%.

So long as you are consistent, you could define the composition in terms of atomic %, or weight fraction, or any appropriate units. Of course the value of the *k* factor would change accordingly. Thus the Cliff–Lorimer equation is the basis for quantitative microanalysis on the AEM. Let's see how we use it in practice.

35.3. PRACTICAL STEPS FOR QUANTITATIVE MICROANALYSIS

First of all, you should try to use only the K_α lines for the measured intensity. (The K_β is combined with K_α if it cannot be resolved.) Use of L or M lines is more difficult because of the many overlapping lines in each family, but may be unavoidable if the K_α lines are too energetic for your Si(Li) detector.

We can break the process down into four accumulation steps:

- Accumulate enough counts in the characteristic peaks, I_A, I_B, etc. As we will see below, for acceptable errors, there should ideally be at least 10^4 counts above background in each peak. While you can't always obtain this number in a reasonable time before specimen drift, damage, or contamination limit your analysis, you should always choose the largest probe size which is consistent with maintaining the desired spatial resolution, so you get most current into your specimen.
- Keep your specimen as close to 0° tilt as possible to minimize spurious effects.
- Orient your specimen so the thin portion of the wedge faces the detector to minimize X-ray absorption (see Section 35.5).
- If the area of the specimen is close to a strong two-beam dynamical-diffraction condition, you should tilt the specimen slightly to kinematical conditions.

Anomalous X-ray generation can occur across bend contours or whenever a diffracted beam is strongly excited. This point is not too critical because we quantify using a ratio technique. If the beam has a large convergence angle, which is usually the case, any diffraction effect is further reduced. We will see in Section 35.8 that under certain conditions there are advantages to be gained from such crystallographic effects.

Having accumulated a spectrum under these conditions, how do you quantify it? All you have to do is measure the peak intensities I_A, I_B, etc., and then determine a value for the k_{AB} factor. To determine the peak intensities you first have to remove the background intensity from the spectrum and then you integrate the peak intensity. In a modern computerized MCA system, both these steps are accomplished by one of several available software routines. There are advantages and disadvantages to each approach, so you should pick the one that is most suited to your problem.

35.3.A. Background Subtraction

We are not very precise in the terminology that we use in the discussion of the X-ray background intensity, so it can be confusing. The "background" refers to the intensity under the characteristic peaks in the spectrum displayed on the MCA screen. Now, as we saw back in Chapter 4, these X-rays are generated by the "bremsstrahlung" or "braking

radiation" process as the beam electrons interact with the coulomb field of the nuclei in the specimen. The intensity distribution of these bremsstrahlung X-rays decreases continuously as the X-ray energy increases, reaching zero at the beam energy. Thus the energy distribution can be described as a "continuum," although, as we've seen, the phenomenon of coherent bremsstrahlung disturbs this continuum.

> Now we tend to use these three terms—"background," "bremsstrahlung," and "continuum"—interchangeably, although strictly speaking they have these specific meanings.

Remember also that the generated bremsstrahlung intensity is modified at energies below about 2 keV by absorption within the detector and the specimen, so we are usually dealing with a background in the spectrum that looks something like Figure 35.1. The best approach to background subtraction depends on whether the region of interest in your spectrum is in this low-energy regime, and if the characteristic peaks you want to measure are close together or isolated.

Window methods. In the most simple case of isolated characteristic peaks superimposed on a slowly varying background, you can easily remove the background intensity by drawing a straight line below the peak and defining the background intensity as that present below the line, as shown in Figure 35.2. So you get the computer first to define a "window" in the spectrum spanning the width

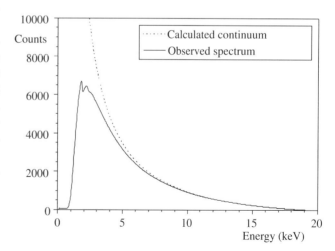

Figure 35.1. The theoretically generated and experimentally observed bremsstrahlung intensity distribution as a function of energy. Both curves are similar until below about 2 keV, when absorption within the specimen and the XEDS system reduces the experimental intensity. Background removal depends on where in the spectrum your characteristic peaks are present.

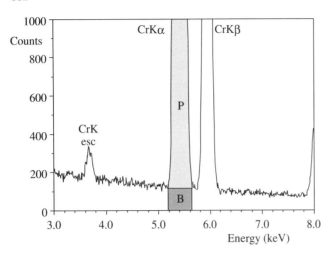

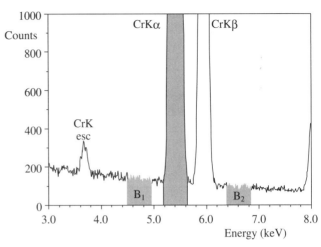

Figure 35.2. The simplest method of estimating the background contribution (*B*) to the intensity in the characteristic peak (*P*); a straight line drawn beneath the Cr K_α peak provides a good estimate if the counting statistics are good, and the bremsstrahlung intensity approximates to a slowly varying linear function of energy. There should be no overlap with any other characteristic peaks, and the peaks should be well above ~2 keV.

Figure 35.3. Background subtraction can be achieved by averaging the bremsstrahlung intensity in two identical windows (B_1, B_2) either side of the characteristic (Cr K_α and K_β) peaks. There should be no overlap with any other characteristic peaks, and the peaks should be well above ~2 keV.

of the peak, and then draw the line between the background intensities in the channels just outside the window. As with all spectral manipulations, this method gives better results with greater intensity levels in the spectrum. The background intensity variation is then less noisy, so it is easier to decide where the peak ends and the background begins. Furthermore, the background intensity variation better approximates to a straight line.

Another similarly primitive approach involves averaging the bremsstrahlung intensity above and below the characteristic peak by integrating the intensity in two identical windows either side of the peak, as shown in Figure 35.3. You then assume that the average of the two intensities equals the background intensity under the peak, so you subtract this average from the total peak intensity. This assumption is reasonable in the higher-energy regions of the spectrum and when the specimen is thin enough so that the bremsstrahlung is not absorbed in the specimen, which would cause a discrete change in intensity under the peak. When you use this approach it is essential to remember the window width you used, because *identical* windows must be used when subtracting the background both in the unknown spectrum and in the spectrum from the known specimen used to determine the *k* factor (see Section 35.4).

> A typical choice of window width is FWHM, but this throws away a substantial amount of the intensity in the peak. FWTM gives better statistics, but incorporates more bremsstrahlung; 1.2 FWHM is the optimum window.

While the two techniques we just described have the advantage of simplicity, you can't always apply them to real specimens because the spectral peaks may overlap. Also, if the peaks lie in the low-energy region of the spectrum where the background is changing rapidly due to absorption, then neither of these two simple methods gives a good estimate of the background and more sophisticated mathematical approaches are required. We'll now discuss these methods.

Modeling the background. The bremsstrahlung intensity distribution can be mathematically modeled, based on the expression developed by Kramers (1923). The number (N_E) of bremsstrahlung photons of energy *E* produced in a given time by a given electron beam is given by Kramers' Law:

$$N_E = KZ\frac{(E_0 - E)}{E} \qquad [35.7]$$

Here, *Z* is the *average* atomic number of the specimen, E_0 is the beam energy in keV, and *E* is the X-ray energy in keV. The factor *K* in Kramers' Law actually takes account of numerous parameters. These include

- Kramers' original constant.
- The collection efficiency of the detector.
- The processing efficiency of the detector.
- The absorption of X-rays within the specimen.

All these terms have to be factored into the computer calculation when you use this method of background modeling. In Figure 35.4, the bremsstrahlung is modeled using the Kramers–Small cross section in the DTSA software (see Section 1.5).

> Be wary when using this approach because Kramers developed his law for bulk specimens. However, the expression is still used in commercial software and seems to do a reasonable job.

A more satisfactory approach, from a scientific standpoint, is to use the modified Bethe–Heitler formula, as discussed by Chapman *et al.* (1984). This formula is explicitly derived for thin foils and high-keV electrons. This model yields an expression for the bremsstrahlung cross section as a function of the X-ray energy (E) and the atomic number (Z) of the specimen. The slow variation of the cross section with Z leads to the possibility of fitting a simple quadratic expression of the form

$$N_E = \left(\frac{a_0}{E} + a_1 + a_2E\right)\varepsilon \qquad [35.8]$$

where a_i are simply fitting parameters and ε is the detector efficiency (plotted back in Figure 32.7). The term is only important when modeling the background below ~1.5 keV, and we discuss it in detail later on in Section 35.4.

    Modeling the spectrum produces a smooth curve fit that describes the shape of the complete spectrum. This approach is particularly valuable if many characteristic peaks are present, since then it is difficult to make local measurements of the background intensity by a window method. Figure 35.4 shows an example of a spectrum containing three adjacent low-energy peaks, with the background intensity estimated underneath all the peaks.

    Filtering out the background. Another mathematical approach to removing the background uses digital filtering. This process makes no attempt to take into account the physics of X-ray production and detection as in Kramers' Law. Rather, it relies on the fact that the characteristic peaks show a rapid variation of intensity as a function of energy (dI/dE is large), while the background exhibits a relatively small dI/dE. This approximation is valid even in the region of the spectrum below ~1.5 keV, where absorption is strong. In the process of digital filtering, the spectrum intensity is "filtered" by convoluting it with another mathematical function. The most common function used is a "top-hat" filter function, so called because of its shape.

> When the top-hat filter is convoluted with the shape of a typical X-ray spectrum, it acts to produce a second-difference spectrum, i.e., d^2I/dE^2.

After the top-hat filter, the background with small dI/dE is transformed to a linear function with a value of zero (thus it is "removed"), while the peaks with large dI/dE, al-

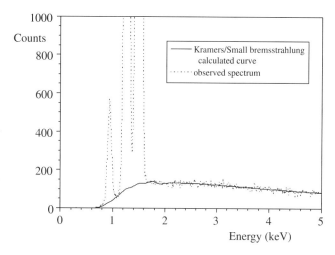

Figure 35.4. The bremsstrahlung intensity modeled using a modified Kramers' Law, which includes the effects of absorption of low-energy X-rays in the specimen and the detector. This method is useful when the spectrum contains overlapping peaks, particularly in the low-energy range, such as the Cu L$_\alpha$ and the Mg and Al K$_\alpha$ lines shown in this spectrum.

though distorted to show negative intensities in some regions, are essentially unchanged as far as the counting statistics are concerned. Figure 35.5A shows schematically the steps required for the filtering process and Figures 35.5B and C show an example of a spectrum before and after digital filtering.

    In summary, you can remove the background by selecting appropriate windows to estimate the intensity under the peak, or use one of two mathematical modeling approaches. The window method is generally good enough if the peaks are isolated and on a linear portion of the background. The mathematical approaches are most useful for multi-element spectra and/or those containing peaks below ~1.5 keV. You should choose the method that gives you the most reproducible results and you must always take care to apply the same process to both the standard and the unknown. After removing the background, the next thing you have to do is integrate the peak intensities I_A, I_B, etc.

35.3.B. Peak Integration

If you used a window method of background estimation, then the peak intensity is obtained simply by subtracting the estimated background intensity from the total intensity in the chosen window. Therefore, if you drew a line under the peak as in Figure 35.2, then the peak intensity is that above the line.

■   If you chose a window of FWHM and averaged the background on either side of the peak, then

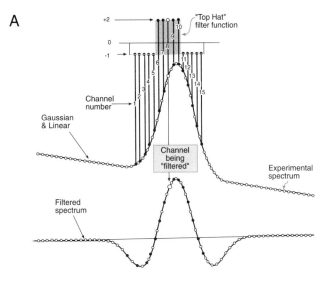

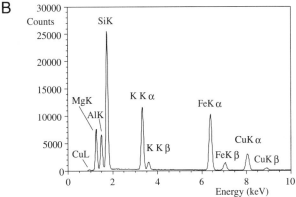

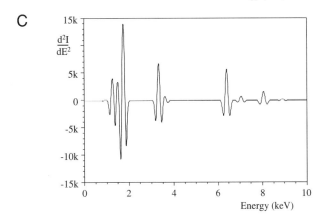

Figure 35.5. (A) Digital filtering involves convolution of a top-hat filter function with the acquired spectrum. To obtain the filtered spectrum, each channel has the top-hat filter applied to it. The channels either side of that being filtered (#8 in this case) are multiplied by the appropriate number in the top-hat function. So channels 1–5 and 11–15 are multiplied by –1 and channels 6–10 by +2. The sum of the multiplications is divided by the total number of channels (15) and allotted to channel #8 in the filtered spectrum at the bottom. The digital filtering process in (A) applied to a spectrum from biotite (B) results in the filtered spectrum (C) in which the background intensity is assigned to zero at all places, and the characteristic peaks remain effectively unchanged.

the average value must be subtracted from the total intensity in the FWHM window.

■ If you used a Kramers' Law fit, the usual method of peak integration is to get the computer to fit the peak with a slightly modified Gaussian, and integrate the total counts in the channels under the Gaussian.

■ If a digital filter was used, you have to compare the peaks with those that were taken previously from standards, digitally filtered, and stored in a "library" in the computer. The library peaks are matched to the experimental peaks via a multiple least-squares fitting procedure and the intensity determined through calculation of the fitting parameters.

Each of these curve-matching processes is rapid. Each can be used to deconvolute overlapping peaks, and each uses all the counts in the peak. The Kramers' fit and the digital filter have much wider applicability than the simple window methods. However, these computer processes are not invariably the best, nor are they without error.

The Gaussian curve fitting must be flexible enough to take into account several variables:

■ The peak width can change as a function of energy or as a function of count rate.

■ The peak "tailing" due to incomplete charge collection can vary.

■ There may be a low-energy background "shelf" and an absorption edge if the specimen is too thick.

The digital-filter approach requires comparison of experimental peaks with library standards, and this means that you have to create a library of stored spectra under conditions that match those liable to be encountered during microanalysis (particularly, similar count rates and dead times). This is a tedious exercise. However, you do get a figure of merit for the "goodness of fit" between the unknown spectrum and the standard. Usually a chi-squared value is given which has no absolute significance, but is a most useful diagnostic tool. Typically, the chi-squared value should be close to unity for a good fit, although a higher value may merely indicate that some unidentified peaks were not accounted for during the matching process. What you have to watch out for is a sudden increase in chi-squared compared with previous values. This indicates that something has changed from your previous analyses. Perhaps your standard is not giving a good fit to the experimental spectrum and either a new library spectrum needs to be gathered or the experimental peak should be looked at carefully. For example, another small peak may be hid-

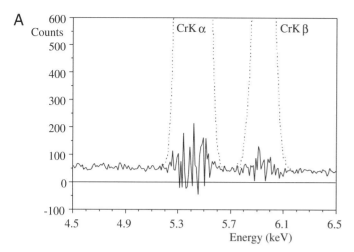

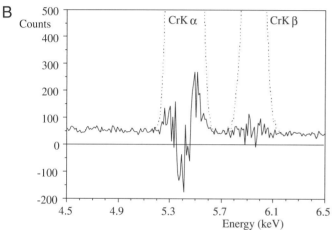

Figure 35.6. (A) A filtered Cr spectrum showing the residual background intensity after the peaks have been removed for integration. The approximately linear residual intensity distribution indicates that the peak matched well with the library standard stored in the computer. (B) A similar filtered spectrum showing the distorted residual counts characteristic of a poor fit with the library standard.

den under the major peak and would need to be deconvoluted from the major peak before integration proceeds. If you suspect a poor fit, you should make the computer display the "residuals," that is, the intensity remaining in the spectrum after the peak has been integrated and removed. As shown in Figure 35.6, you can easily see if a good fit was made (Figure 35.6A) or if the library peak and the experimental peak do not match well (Figure 35.6B).

> The point we are making is that any of the above methods is valid for obtaining values of the peak intensity. They should all result in the same answer when used to quantify an unknown spectrum, so long as you apply the same method consistently to both the standard and the unknown.

The next step is to insert the values of the peak intensities in the Cliff–Lorimer ratio equation and know the correct value of the k factor. So we now need to discuss the various ways to obtain k_{AB}.

35.4. DETERMINING *k* FACTORS

Remember that the k factor is not a constant. It is a sensitivity factor that will vary not only with the X-ray detector, the microscope, and the microanalysis conditions, but also with your choice of background-subtraction and peak-integration methods. So values of k factors can be sensibly compared *only* when they were obtained under identical conditions. We will return to this point at the end of this section when we look at various sets of k factors published in the literature. There are two ways you can determine k factors:

■ Experimental determination using standards.
■ Calculation from first principles.

The first method is slow and laborious but gives the most accurate values. The second method is quick and painless but the results are less reliable.

35.4.A. Experimental Determination of k_{AB}

If you have a thin specimen of known composition, C_A, C_B, etc., then all you have to do, in principle, is place that specimen in the microscope, generate a spectrum, obtain values of I_A, I_B, etc., and insert those values in the Cliff–Lorimer equation (equation 35.2). Since you know C_A and C_B, the only unknown is k_{AB}. However, there are several precautions that you must take before this procedure can be used:

■ The standard must be a well-characterized specimen, and it is usually best if it is *single phase*.
■ The standard must be capable of being thinned to electron transparency. Ideally, when the specimen is thin there should be no significant absorption or fluorescence of the X-rays from the elements A, B, etc. that you wish to analyze.
■ You must be sure that the thinning process did not induce any chemical changes (this is discussed in some detail in Chapter 10).
■ It must be possible to select thin regions that are characteristic of the bulk.
■ You must be sure that the thin foil is stable under the electron beam at the voltage you intend to use for microanalysis.

This last point may often be the limiting factor in your choice of standards because, as we saw in Section 4.6, you have to take care to avoid not only direct knock-on damage, but also sputtering effects which occur at voltages substantially below the threshold for direct atom displacement. Obviously, both these problems become greater as the beam voltage increases.

The National Institute of Standards and Technology (NIST) has issued a thin standard containing the elements Mg, Si, Ca, Fe, and O (SRM #2063). Unfortunately, X-rays from the lighter elements in this standard film are absorbed significantly in the film, and also in the detector, and so a correction to the measured k factor is necessary. In fact, there are no generally accepted standards that meet all the above criteria for ideal k-factor determination. It is best to use your own judgment in this respect, and also make use of the knowledge gained in previous k-factor studies. The original work of Cliff and Lorimer (1975) was based on crushed mineral standards. Their approach has two advantages:

■ Crushing does not affect the chemistry; the stoichiometry is well known.
■ Minerals contain Si, thus permitting the creation of a whole series of k_{ASi} factors.

The drawbacks are that the mineral samples often contain more than one phase, or may be naturally nonstoichiometric. Clearly, some prior knowledge of the mineralogy of the sample is essential in order to be able to select the right spectrum to use as a standard. Also, Si K_α X-rays at ~1.74 keV are liable to be absorbed in the XEDS detector, so there may be a systematic difference in k factors determined with different detectors. Finally, silicate minerals often exhibit radiolysis, i.e., chemical changes due to beam-induced breaking of bonds.

Several alternative approaches that attempt to avoid the problems with k_{ASi} have been proposed:

■ Wood $et\ al.$ (1984) generated a series of k_{AFe} factors to overcome the Si absorption and the beam sensitivity problems.
■ Graham and Steeds (1984) used crystallized microdroplets.
■ Sheridan (1989) demonstrated the value of the NIST multi-element glasses.

In all cases the bulk chemistry has to be determined by some acceptable technique, such as EPMA, atomic absorption spectroscopy, or wet chemistry. Since all these techniques analyze relatively large volumes of material, it is best that the standard be single phase. However, because none of these techniques can determine if the specimen is homogeneous on a submicron scale, the only way to find out the level of homogeneity is to carry out many analyses within the AEM to confirm that any variation in your answer is within the expected X-ray statistical fluctuations. A typical k-factor determination therefore involves taking many spectra from different parts of the thin-foil standard to check both the homogeneity and the stability of the specimen. Each spectrum should contain sufficient counts in the peaks of interest to ensure that the errors in the k-factor determination are at least less than ±5% relative and, if possible, less than ± 3%. So, now we need to consider the errors associated with the X-ray spectra.

35.4.B. Errors in Quantification; the Statistics

An unfortunate aspect of the simple Cliff–Lorimer ratio equation is that it has relatively large errors associated with it. The thin foil that removes the problems of absorption and fluorescence usually results in relatively few X-ray photons per incident electron, compared with bulk specimens. This effect is compounded by the small collection angle of the XEDS detector. The end result is that poor counting statistics are the primary source of error in the quantification. The best way you can limit these errors is to use higher-brightness sources, large electron probes, and thicker specimens, unless absorption is a problem, or spatial resolution is paramount. In any case you should be prepared to count for a long time, assuming that specimen drift and/or contamination don't compromise the data.

> Experimental results show that the X-ray counts in the spectrum obey Gaussian statistics. Hence we can apply simple statistics to deduce the accuracy of any quantification.

The rest of this section is purely statistics. If you know it, then jump ahead.

Given that our characteristic peak is Gaussian, then the standard deviation σ is obtained from

$$\sigma = N^{\frac{1}{2}} \qquad [35.9]$$

where N is the number of counts in the peak above the background. For a single measurement there is a 67% chance that the value of N will be within 1σ of the true value of N. This chance increases to 95% for 2σ and 99.7% for 3σ. If we use the most stringent condition, then the relative error in any single measurement is

$$\text{Relative Error} = \frac{3N^{\frac{1}{2}}}{N} 100\% \qquad [35.10]$$

Clearly the error decreases as N increases, and hence the emphasis throughout this chapter is on the need to maximize the X-ray counts gathered in your spectra. Since the Cliff–Lorimer equation uses an intensity ratio, we can get a quick estimate of the error by summing the errors in I_A, I_B, and k_{AB} to give the total error in the composition ratio C_A/C_B.

Summing the errors in fact gives an overestimate of the error and, strictly speaking, we should add the relative standard deviations in quadrature using the expression

$$\left(\frac{\sigma_C}{C_B/C_B}\right)^2 = \left(\frac{\sigma_{k_{AB}}}{k_{AB}}\right)^2 + \left(\frac{\sigma_{I_A}}{I_A}\right)^2 + \left(\frac{\sigma_{I_B}}{I_B}\right)^2 \qquad [35.11]$$

So we can determine the error for each datum point in this manner. If we are determining the composition of a single phase, for example during the determination of a k factor, then we can reduce the error by combining the results from n different measurements of the intensity ratio I_A/I_B. The total absolute error in I_A/I_B at a given confidence limit is obtained using the student "t" distribution. For example, in this approach the error is given by

$$\text{Absolute Error} = \frac{(t_{95})^{n-1} S}{n^{\frac{1}{2}}} \qquad [35.12]$$

where t_{95}^{n-1} is the student "t" value at the 95% confidence limit for n measurements of k_{AB} (see any statistics text for a list of student "t" values, e.g., Owen 1962). Obviously, you could choose a lower or higher confidence level. Here, S is the standard deviation for n measurements of the intensity, N_i, which on average contain $\bar{N}_i$ counts.

$$S = \left(\sum_{i=1}^{n} \frac{(N_i - \bar{N}_i)^2}{n-1}\right)^{1/2} \qquad [35.13]$$

Hence by increasing the number of measurements n, you can reduce the absolute error in k_{AB}. With enough measurements and a good homogeneous specimen you can reduce the errors in the value of k_{AB} to $\pm 1\%$, as we will see in the example below. However, remember that this figure must be added to the errors in I_A and I_B. From equation 35.9 it is easy to determine that if we accumulate 10,000 counts in the peak for element A, then the error at the 99% confidence limit is [3 (10,000)$^{1/2}$ / 10,000] × 100%, which is ~3%. A similar value for I_B gives a total error in C_A/C_B of ~±4.5%, using equation 35.11. If you take the time to accumulate 100,000 counts for I_A and I_B, the total error is reduced to ~±1.7%, which represents about the best accuracy that can be expected for quantitative analysis in the AEM. It is appropriate here to go through an illustration of a k_{AB} determination using actual experimental data.

Before deciding that a particular specimen is suitable, it should be checked for its level of homogeneity, and there is a well-established criterion for this. If we take the average value N of many composition determinations, and all the data points fall within $\pm 3(N)^{1/2}$ of N, then the sample is homogeneous. In other words, this is our *definition of homogeneous*. There are more rigorous definitions, but the general level of accuracy in thin-foil microanalysis is such that there is no need to be concerned about them.

Example

A homogenized thin foil of Cu-Mn solid solution was used to determine k_{CuMn}. The sample was first analyzed by EPMA and found to be 96.64 wt.% Cu and 3.36 wt.% Mn. Since our accuracy is increased by collecting many spectra, a total of 30 were accumulated ($n = 30$ in equation 35.13). In a typical spectrum, the Cu K$_\alpha$ peak contained 271,500 counts above background and the Mn K$_\alpha$ peak contained 10,800 counts. So if we insert these data into the Cliff–Lorimer equation we get

$$\frac{96.64}{3.36} = k_{CuMn} \frac{271,500}{10,800}$$

$$k_{CuMn} = 1.14$$

To determine an error on this value of the k factor, equation 35.12 must be used. The student "t" analysis of the k factors from the other 29 spectra gives an error of ± 0.01 for a 95% confidence limit. This error of about $\pm 1\%$ relative is about the best that can be achieved using the experimental approach to k-factor determination.

Tables 35.1 and 35.2 summarize many of the available k-factor data in the published literature. You should go and read the original papers, particularly if you want to find out what standards and what conditions were used in their determination.

35.4.C. Calculating k_{AB}

While it is clear that many of the values in the tables are very similar, the differences cannot be accounted for by X-ray statistics alone. Some of the differences arise due to the choice of standard and the reproducibility of the standard. Other differences arise because the data were obtained under different conditions, such as different peak-integration routines. Therefore, the point made at the beginning of this section is worth repeating.

> The k factors are *not* standards, but sensitivity factors.

Table 35.1. Experimentally Determined k_{ASi} and k_{AFe} Factors for K_α X-ray Lines[a]

Element (A)	k_{ASi} (1) 100 kV	k_{ASi} (2) 100 kV	k_{ASi} (3) 120 kV	k_{ASi} (4) 80 kV	k_{ASi} (5) 100 kV	k_{ASi} (5) 200 kV	k_{AFe} (6) 120 kV	k_{ASi} (7) 200 kV
Na	5.77	3.2	3.57 ± 0.21	2.8 ± 0.1	2.17	2.42		3.97 ± 2.32
Mg	2.07 ± 0.1	1.6	1.49 ± 0.007	1.7 ± 0.1	1.44	1.43	1.02 ± 0.03	1.81 ± 0.18
Al	1.42 ± 0.1	1.2	1.12 ± 0.03	1.15 ± 0.05			0.86 ± 0.04	1.25 ± 0.16
Si	1.0	1.0	1.0	1.0	1.0	1.0	0.76 ± 0.004	1.00
P			0.99 ± 0.016				0.77 ± 0.005	1.04 ± 0.12
S			1.08 ± 0.05		1.008	0.989	0.83 ± 0.03	1.06 ± 0.12
Cl					0.994	0.964		1.06 ± 0.30
K		1.03	1.12 ± 0.27	1.14 ± 0.1			0.86 ± 0.014	1.21 ± 0.20
Ca	1.0 ± 0.07	1.06	1.15 ± 0.02	1.13 ± 0.07			0.88 ± 0.005	1.05 ± 0.10
Ti	1.08 ± 0.07	1.12	1.12 ± 0.046				0.86 ± 0.02	1.14 ± 0.08
V	1.13 ± 0.07			1.3 ± 0.15				1.16 ± 0.16
Cr	1.17 ± 0.07	1.18	1.46 ± 0.03				0.90 ± 0.006	
Mn	1.22 ± 0.07	1.24	1.34 ± 0.04				1.04 ± 0.025	1.24 ± 0.18
Fe	1.27 ± 0.07	1.30	1.30 ± 0.03	1.48 ± 0.1			1.0	1.35 ± 0.16
Co							0.98 ± 0.06	1.41 ± 0.20
Ni	1.47 ± 0.07	1.48	1.67 ± 0.06				1.07 ± 0.006	
Cu	1.58 ± 0.07	1.60	1.59 ± 0.05		1.72	1.50	1.17 ± 0.03	1.51 ± 0.40
Zn	1.68 ± 0.07				1.74	1.55	1.19 ± 0.04	1.63 ± 0.28
Ge	1.92							1.91 ± 0.54
Zr								3.62 ± 0.56
Nb							2.14 ± 0.06	
Mo	4.3		4.95 ± 0.17				3.8 ± 0.09	
Ag	8.49		12.4 ± 0.63				9.52 ± 0.07	6.26 ± 1.50
Cd	10.6				9.47	6.2		
In								7.99 ± 1.80
Sn	10.6							8.98 ± 1.48
Ba					29.3	17.6		21.6 ± 2.6

Table 35.2. Experimentally Determined k_{ASi} and k_{AFe} Factors for L_α X-ray Lines[a]

Element (A)	k_{ASi} (8) 100 kV	k_{ASi} (5) 100 kV	k_{ASi} (5) 200 kV	k_{ASi} (9) 100 kV	k_{AFe} (6) 120 kV	k_{ASi} (7) 200 kV
Cu		8.76	12.2			
Zn		6.53	6.5			8.09 ± 0.80
Ge						4.22 ± 1.48
As						3.60 ± 0.72
Se						3.47 ± 1.11
Sr					1.21 ± 0.06	
Zr					1.35 ± 0.1	2.85 ± 0.40
Nb					0.9 ± 0.06	
Mo				2.0		
Ag	2.32 ± 0.2				1.18 ± 0.06	2.80 ± 1.19
In					2.21 ± 0.07	2.86 ± 0.71
Cd		2.92	2.75			
Sn	3.07 ± 0.2					
Ba		3.38	2.94			3.36 ± 0.58
Ce				1.4		
Sn	3.1 ± 0.2			1.3		
W	3.11 ± 0.2			1.8		3.97 ± 1.12
Au	4.19 ± 0.2	4.64	3.93		3.1 ± 0.09	4.93 ± 2.03
Pb	5.3 ± 0.2	4.85	4.24	2.8		5.14 ± 0.89

[a]Sources: (1) Cliff and Lorimer (1975), (2) Wood *et al.* (1981), (3) Lorimer *et al.* (1977), (4) McGill and Hubbard (1981), (5) Schreiber and Wims (1981), (6) Wood *et al.* (1984), (7) Sheridan (1989), (8) Goldstein *et al.* (1977), (9) Sprys and Short (1976).

The only conditions under which you can expect the k factors obtained on different AEMs to be identical are if you use the *same* standard at the *same* accelerating voltage, *same* detector configuration, and *same* peak-integration and background-subtraction routines. Even then there will be differences if one or more of the measured X-ray lines is not gathered by the detector with 100% efficiency; the X-ray may be either absorbed by the detector or it may be too energetic and pass straight through the detector.

You may not be able to obtain a suitable standard. For example, you might be working in a system in which no stoichiometric phases exist or accuracy might not be critical, but you need a quick analysis. Then you can calculate an approximate k factor. The programs necessary to calculate k_{AB} are stored in the computer and will give a value of k in a fraction of a second. The calculated value should be accurate to within ±20% relative. Often, this level of accuracy is all you need to draw a sensible conclusion concerning the problem.

> Calculating k factors is the recommended approach when a quick answer is required and accuracy is not essential.

We will derive the expression for calculating the k factor from first principles, starting in a manner similar to the development of the expressions for the analysis of bulk samples in the EPMA. The derivation gives a good illustration of the relationship between bulk and thin-film microanalysis, and provides insight into the details of X-ray interactions with solids. In addition, the equations will provide us with the necessary grounding to pursue the problems of absorption and fluorescence in thin foils, when they occur. A full discussion of this derivation is given in the paper by Williams and Goldstein (1991). If you don't need to know the details of this derivation, you may wish to move on to the final expression given in equation 35.23.

The intensity of the *generated* X-ray emission from element A in the specimen, I_A^{Spec*}, is

$$I_A^{Spec*} = \Phi_A^{\Delta \rho t} \int_0^\infty \varphi(\rho t)\, e^{-\chi \rho t} \left(1 + \delta_A\right) d(\rho t) \qquad [35.14]$$

- The term $\Phi_A^{\Delta \rho t}$ is the X-ray emission (in cps) generated from element A in an isolated thin film of the specimen with mass thickness $\Delta \rho t$; the thickness of this isolated film is Δ and its mass thickness is ρt (it is *not* the change in ρt).
- The term $\varphi(\rho t)$ is the depth distribution of X-ray production. We define it as the ratio of the X-ray emission from a layer of element A of

thickness $\Delta \rho t$ at a depth t in the specimen to the X-ray emission from an identical, but isolated, film.

- The expression $e^{-\chi \rho t}$ accounts for X-ray absorption in the specimen, where χ is defined as

$$\chi = \frac{\mu}{\rho}\Big]_{Spec}^A \csc \alpha \qquad [35.15]$$

- The term $\mu/\rho]_{Spec}^A$ is the mass absorption coefficient for X-rays from element A in the specimen and α is the X-ray take-off angle.

X-rays from element A may also be fluoresced by other characteristic X-rays emerging from the specimen. The fluorescence contribution to the generated intensity is $(1+\delta_A)$. We can write an expression for the intensity of X-rays from an isolated thin film within a specimen as

$$\Phi_A^{\Delta \rho t} = N\left(\frac{Q\omega a}{A}\right)_A C_A \Delta \rho t \qquad [35.16]$$

where N is Avogadro's number. The subscript A denotes the element A in each case, Q_A is the ionization cross section, ω_A is the fluorescence yield for the characteristic X-rays, A_A is the atomic weight and C_A is the weight fraction of the element.

> The use of the weight fraction rather than the atomic fraction is an anomaly which has persisted from the earliest days of microanalysis, when it was thought by Castaing that an atomic number correction was not required.

The remaining term "a" is the relative transition probability. This term takes account of the fact that when a K-shell electron is ionized, the atom will return to ground state through the emission of either a K_α or K_β X-ray. The term "a" in equation 35.14 in this case would be given by

$$a = \frac{I(K_\alpha)}{I(K_\alpha + K_\beta)} \qquad [35.17]$$

You may remember that we listed the relative "weights" of the X-ray lines in Table 4.1. We can easily apply equation 35.14 to thin-film specimens. For such specimens we can then make some simplifications:

- Assume that the electrons only lose a small fraction of their energy in traversing the specimen. Therefore, Q_A is taken as a constant, and evaluated for the incident beam energy E_0.

■ Limit the integral in equation 35.12 to the foil thickness t.

Thus if we substitute equation 35.16 into equation 35.14, we find

$$I_A^{Spec} = N\left(\frac{Q\omega a}{A}\right)_A C_A \Delta\rho t \int_0^t \varphi_A(\rho t)\, e^{-\chi\rho t}(1+\delta_A)\, d(\rho t) \quad [35.18]$$

The Cliff–Lorimer equation assumes that we can measure two characteristic X-ray intensities simultaneously and so we can ratio two equations like equation 35.18, cancel N and $\Delta(\rho t)$, and rewrite them thus

$$\frac{I_A^{Spec\,*}}{I_B^{Spec\,*}} = \frac{C_A \dfrac{Q_A\omega_A a_A}{A_A} \displaystyle\int_0^t \varphi_A(\rho t)\, e^{-\chi\rho t}(1+\delta_A)\, d(\rho t)}{C_B \dfrac{Q_B\omega_B a_B}{A_B} \displaystyle\int_0^t \varphi_B(\rho t)\, e^{-\chi\rho t}(1+\delta_B)\, d(\rho t)} \quad [35.19]$$

This equation can be conveniently shortened to

$$\frac{I_A}{I_B} = \frac{C_A}{C_B}(ZAF) \quad [35.20]$$

where Z, A, and F stand for the atomic-number, absorption, and fluorescence corrections, respectively. Now remember that the Cliff–Lorimer equation (equation 35.2) assumes that A and F are negligible in a thin foil. We therefore rearrange equation 35.20 to look like equation 35.2

$$\frac{C_A}{C_B} = \frac{1}{Z}\frac{I_A}{I_B} \quad [35.21]$$

By comparison of the two equations (35.2 and 35.19) we can write an expression for k_{AB}

$$k_{AB} = \frac{1}{Z} = \frac{(Q\omega a)_B A_A}{(Q\omega a)_A A_B} \quad [35.22]$$

Thus, as we mentioned at the start of the discussion on quantification, the Cliff–Lorimer k factor for thin-foil analysis is related to the atomic-number correction factor (Z) for bulk specimen microanalysis. From equation 35.22 we can easily see which experimental factors determine the value of k.

■ Obviously, the accelerating voltage is a variable since Q is strongly affected by the kV.
■ The atomic number affects ω, A, and a.
■ The choice of peak-integration method will also affect a.

Therefore, in order to calculate and compare different k factors, it is imperative to define these conditions very clearly, as we have taken pains to emphasize.

Equation 35.20 assumes that equal fractions of the X-rays generated by elements A and B are collected and processed by the detector. This assumption will only be true if the same detector is used and the X-rays are neither strongly absorbed nor pass completely through the detector. However, as we have already seen in Chapter 32, X-rays below ~1.5 keV are absorbed significantly by the Be window and X-rays above ~20 keV pass through a 3-mm Si detector with ease. Under these circumstances it is necessary to modify the k-factor expression, equation 35.22, in the following manner

$$k_{AB} = \frac{1}{Z} = \frac{(Q\omega a)_A\, A_B\, \varepsilon_A}{(Q\omega a)_B\, A_A\, \varepsilon_B} \quad [35.23]$$

The symbol ε represents simply a detector-efficiency term (plotted back in Figure 32.7) that we can write as follows

$$\varepsilon_A = \exp\left(-\frac{\mu}{\rho}\Big]_{Be}^A \rho_{Be} t_{Be}\right) \exp\left(-\frac{\mu}{\rho}\Big]_{Au}^A \rho_{Au} t_{Au}\right)$$

$$\cdot \exp\left(-\frac{\mu}{\rho}\Big]_{Si}^A \rho_{Si} t_{Si}\right)\left\{1 - \exp\left(-\frac{\mu}{\rho}\Big]_{Si}^A \rho_{Si} t'_{Si}\right)\right\} \quad [35.24]$$

Here, the mass absorption coefficient for the X-rays from element A are required for the Be window, the Au (or other) contact layer, the Si dead layer, and the Si intrinsic region (thickness t'). We can also write a similar expression for an IG detector. These various detector parameters each have density ρ (available from standard elemental density tables) and thickness t. Typical values of t for each part of the detector were discussed in Chapter 32. The first three terms thus account for absorption of weak X-rays passing through the Be window, the Au contact layer, and the Si dead layer before entering the detector. The last term adjusts the k factor for X-rays that *do not* deposit their energy in the active region of the detector which has density ρ and thickness t'.

While equations 35.23 and 35.24 look simple for a computer to solve, the values that have to be inserted in the equations for the various terms are not always well known, or cannot be measured accurately. For example, we do not know the best value of Q for many elements in the range of voltages typically used in the AEM (100–400 kV). There are considerable differences of opinion in the literature concerning the best way to choose a value for Q. The two major approaches used are:

■ Assume various empirical parameterization processes (e.g., Powell 1976).
■ Interpolate values of Q to give the best fit to experimental k factors (Williams *et al.* 1984).

Table 35.3a. Calculated k_{AFe} Factors for K Lines Using Different Theoretical Cross Sections[a]

Element A	k_{MM}	k_{GC}	k_P	k_{BP}	k_{SW}	k_Z
Na	1.42	1.34	1.26	1.45	1.17	1.09
Mg	1.043	0.954	0.898	1.03	0.836	0.793
Al	0.893	0.882	0.777	0.877	0.723	0.696
Si	0.781	0.723	0.687	0.769	0.638	0.623
P	0.813	0.759	0.723	0.803	0.671	0.663
S	0.827	0.776	0.743	0.817	0.688	0.689
K	0.814	0.779	0.755	0.807	0.701	0.722
Ca	0.804	0.774	0.753	0.788	0.702	0.727
Ti	0.892	0.869	0.853	0.888	0.807	0.835
Cr	0.938	0.925	0.917	0.936	0.887	0.909
Mn	0.98	0.974	0.970	0.979	0.953	0.965
Fe	1.0	1.0	1.0	1.0	1.0	1.0
Co	1.063	1.069	1.074	1.066	1.096	1.079
Ni	1.071	1.085	1.096	1.074	1.143	1.23
Cu	1.185	1.209	1.227	1.19	1.31	1.24
Zn	1.245	1.278	1.305	1.255	1.44	1.32
Mo	3.13	3.52	3.88	3.27	3.84	3.97
Ag	4.58	5.41	6.23	4.91	5.93	6.28

The other major variable in equation 35.24 is the Be window thickness, which is nominally 7.5 μm but in practice may be substantially thicker. Tables 35.3a and b list calculated k factors obtained using various expressions for Q. As you can see, the value of k may easily vary by >±10%, particularly for the lighter elements and the heavier elements. This variation is due to the uncertainties in the detector-efficiency terms in equation 35.22. The values of k_{AB} for the L lines are even less accurate than for the K lines, mainly because the values of Q for the L lines are somewhat speculative. There are no data available for calculated k factors for M lines. Under these circumstances, experimental determination is the only approach. This point again emphasizes the advantages of K-line analysis where possible. When the heavy elements are being studied, the L or M

lines, which may be the strongest in a spectrum from a Si(Li) detector, will undoubtedly give rise to greater errors than the K lines, which may only be detectable with an IG system.

The combination of uncertainties in Q and in the detector parameters is the reason why calculated k factors are not very accurate, usually no better than ±10–20% relative. The computer system attached to the AEM will have predetermined values of all the terms in equations 35.21 and 35.22 stored in its memory. You don't usually have control over which particular parameters are being used. However, you should at least ask the manufacturer to list the sources of the values of Q, ω, and a in the computer. You should then carry out a cross-check calculation with a known specimen to ensure that the calculated k factor gives the correct answer.

Table 35.3b. Calculated k_{AFe} Factors for L Lines Using Different Theoretical Cross Sections[a]

Element	k_{MM}	k_P	k_{HP}	k_{SW}	k_Z
Sr[a]	1.73	1.33	1.32	1.64	1.39
Zr[a]	1.62	1.26	1.24	1.51	1.33
Nb[a]	1.54	1.21	1.18	1.43	1.28
Ag[a]	1.43	1.16	1.09	1.26	1.26
Sn	2.55	2.09	1.93	2.21	2.30
Ba	2.97	2.52	2.25	2.49	2.83
W	3.59	3.37	2.68	2.80	3.88
Au	3.94	3.84	2.94	3.05	4.43
Pb	4.34	4.31	3.o5	3.34	4.97

> If you replace or service a detector, which is not an unusual occurrence on an AEM, then the new detector parameters must be inserted into the software.

We cannot recommend a "best" set of values for Q, ω and a, but the values of Q given by Powell (1976), ω by Bambynek *et al.* (1972), and a by Schreiber and Wims (1982) have been used in the past. Also, we can't give you specific detector parameters, so you should obtain an estimate from the manufacturer. The values of μ/ρ which we recommend are those determined by Heinrich (1986), although there is still considerable uncertainty in the mass absorption coefficients for the low-energy X-rays from the light elements. If

[a]k factors use the L intensity from the L_α and L_β lines. MM = Mott–Massey; GC = Green–Cosslett; P = Powell; BP = Brown–Powell; SW = Schreiber–Wims; Z = Zaluzec.

you use the DTSA program from NIST (see Section 1.5), you may find that it predicts a worse value.

> Remember that the problem is that all software packages use preset values in their calculations and these may vary from package to package.

Figures 35.7A and B show a comparison of the two methods of k-factor determination. The experimental data are shown as individual points with error bars, and the solid lines represent the range of calculated k factors, depending on the particular value of Q used in equation 35.21. The relatively large errors possible in the calculated k factors are clearly seen, and comparison of the K-line data in Figure 35.7A with the L-line data in Figure 35.7B again emphasizes the advantages of using K lines for the

analysis where possible. Similar data for M lines are almost nonexistent, but data for the K lines from the heavier elements will become more common if IG detectors are more widely used.

We can summarize the k-factor approach to microanalysis in the following way:

- The Cliff–Lorimer equation has the virtue of simplicity. All you have to do is specify all the variables and treat the standard and unknown in an identical manner.
- You are better off calculating k_{AB} if you prefer speed to accuracy. Experimental determination is best if you wish to have a known level of confidence in the numbers that you produce.

The point at which the simple Cliff–Lorimer approach breaks down is when the thin-foil criterion is invalid. Absorption is far more common than fluorescence in thin foils. You must be wary of absorption when you have X-ray lines in your spectrum that differ in energy by >5–10 keV, particularly if any are light-element X-rays. To understand why this is so, we must investigate the absorption correction factor.

35.5. ABSORPTION CORRECTION

Preferential absorption of the X-rays from one of the elements in your specimen means that the detected X-ray intensity will be less than the generated intensity and so C_A is no longer simply proportional to I_A. So you have to modify your k factor to take into account the reduction in I_A. This is a problem if your specimen is too thick, or if one or more of the characteristic X-rays has an energy less than ~1 keV (i.e., light element analysis). If we define k_{AB} as the true sensitivity factor when the specimen thickness $t = 0$, then the effective sensitivity factor for a specimen in which absorption occurs is given by k_{AB}^* where

$$k_{AB}^* = k_{AB}(ACF) \qquad [35.25]$$

The absorption correction factor (ACF) is the A term in equation 35.20 and this can be written out fully, incorporating the expression for χ from equation 35.15, to give

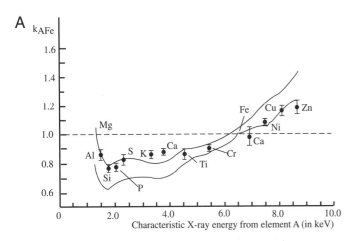

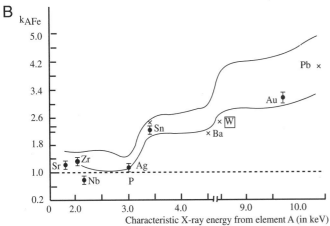

$$ACF = \frac{\displaystyle\int_0^t \left\{ \varphi_B(\rho t)\, e^{-\left(\frac{\mu}{\rho}\Big|_{Spec}^B \rho t\, cosec\, \alpha\right)} \right\} d(\rho t)}{\displaystyle\int_0^t \left\{ \varphi_A(\rho t)\, e^{-\left(\frac{\mu}{\rho}\Big|_{Spec}^A \rho t\, cosec\, \alpha\right)} \right\} d(\rho t)} \qquad [35.26]$$

Figure 35.7. (A) Experimental k_{AFe} factors for the K_α X-rays from a range of elements A with respect to Fe. The solid lines represent the spread of calculated k factors using different values for the ionization cross section. (B) Similar data for L_α lines from relatively high-Z elements. The errors in the calculated values of k are large, reflecting the uncertainties in L-line ionization cross sections.

In this expression $\mu/\rho]^A_{Spec}$ is the mass absorption coefficient of X-rays from element A in the specimen, α is the detector take-off angle, ρ is the density of the specimen, and t is the thickness. Since the units of μ/ρ are usually cm^2/gm, be sure to use ρ in gm/cm^3 and t in cm, rather than SI units. Obviously, the value of the ACF is unity when no absorption occurs. Typically, if the ACF is >10% then the absorption is significant, since 10% accuracy is routinely attainable in quantitative microanalysis using experimental k factors. However, accuracy better than 10% can be obtained if you decide what constitutes a "significant" level of absorption for the problem at hand and the accuracy required of the data. Let's now look at each of the terms and the problems associated with determining their value.

Again, we recommend that you use the values of μ/ρ given by Heinrich (1986). The value of μ/ρ for a particular X-ray (e.g., from element A) within the specimen is the sum of the mass absorption coefficients for each element times the weight fraction of that element, so

$$\frac{\mu}{\rho}\bigg|^A_{Spec} = \sum_i \left(\frac{C_i \mu}{\rho}\bigg|^A_i \right) \qquad [35.27]$$

where C_i is the fractional concentration of element i in the specimen such that

$$\sum_i C_i = 1 \qquad [35.28]$$

The absorption of X-rays from element A by all elements i in the specimen is summed, including self absorption by element A itself, absorption by elements that may not be of interest in the experiment, and even by materials whose X-rays might not be detectable.

An example of this phenomenon occurs when Mg is being quantified in homogeneous NiO-MgO. The Mg K_α X-rays will be absorbed by oxygen, even if the O K_α X-ray is not of interest or cannot be detected because a Be window detector is being used. This effect is shown in Figure 35.8, which shows an increase in the intensity ratio (Ni K_α/Mg K_α) as a function of thickness due to the increased absorption of the Mg K_α X-rays. (Absorption appears in an exponential term.) If we correct for the absorption by Ni, the slope of the line is reduced, but only when the effects of absorption by oxygen are taken into account does the slope become zero, as it should be for a homogeneous specimen (Bender *et al.* 1980).

In equation 35.26, the depth distribution of X-ray production $\varphi(\rho t)$ is assumed to be a constant and equal to

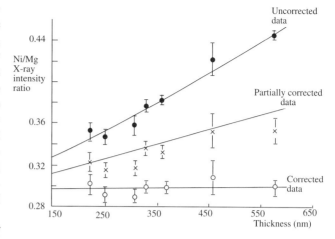

Figure 35.8. The upper curve shows the raw Ni K_α/Mg K_α intensity ratio as a function of thickness in a homogeneous sample of NiO-MgO. The slope indicates strong absorption of Mg K_α X-rays. The middle curve shows the effect of correcting for absorption of the Mg K_α line by Ni and the bottom line shows the effect of a further correction for absorption of the Mg K_α line by O to give the expected horizontal line.

unity. That is, a uniform distribution of X-rays is generated at all depths throughout the foil. This is a reasonable first approximation in thin foils, but in bulk specimens $\varphi(\rho t)$ is a strong function of t. Depending on the thickness of the foil, it is possible that this assumption may be the limiting factor in the accuracy of the absorption correction, but for most thin foils, particularly if Z is < 30, variations in $\varphi(\rho t)$ can be ignored. If $\varphi(\rho t)$ does affect the absorption correction, then it will result in a slight overcompensation for the effects of absorption, which will get worse as the thickness increases.

The measurement of $\varphi(\rho t)$ for bulk specimens is a well-established procedure. The few studies in thin specimens show an increase in $\varphi(\rho t)$ with specimen thickness, although the increase is no more than about 5% in foil thicknesses of <300 nm. Therefore, the assumption that $\varphi(\rho t)$ equals unity does indeed appear reasonable. The fact that we use a ratio of the two $\varphi(\rho t)$ terms in the absorption equation also helps to minimize the effects of this assumption.

We assume that $\varphi(\rho t)$ equals unity. Then we can simply use equation 35.26 to give

$$ACF = \left(\frac{\frac{\mu}{\rho}\Big|^A_{Spec}}{\frac{\mu}{\rho}\Big|^B_{Spec}} \right) \left(\frac{1 - e^{-\left(\frac{\mu}{\rho}\big]^B_{Spec} \rho t \, cosec\, \alpha \right)}}{1 - e^{-\left(\frac{\mu}{\rho}\big]^A_{Spec} \rho t \, cosec\, \alpha \right)}} \right) \qquad [35.29]$$

So we still need to know the values of ρ and t for our specimens.

The density of the specimen (ρ) can be estimated if you know the unit-cell dimensions, e.g., from convergent-beam electron diffraction

$$\rho = \frac{n\,A}{V\,N} \qquad [35.30]$$

where n is the number of atoms of average atomic weight A in a unit cell of volume V, and N is Avogadro's number.

The absorption path length (t') is a major variable in the absorption correction. Fortunately, it is also the one over which you, the operator, have the most control. In the simplest case of a parallel-sided thin foil of thickness t at 0° tilt, the absorption path length, as shown in Figure 35.9, is given by

$$t' = t\,\mathrm{cosec}\,\alpha \qquad [35.31]$$

where α is the detector take-off angle. To minimize this factor, it is obvious that your specimen should be as thin as possible and the value of α as high as possible. There are many ways to determine the foil thickness which we have discussed at various points in this text; they are summarized in Section 36.6. No method is universally applicable, and few are either easy or accurate. The value of α with the specimen at 0° tilt is fixed by the design geometry of the stage and the only way to vary α is by tilting the specimen. As we have seen, there are good reasons not to tilt the specimen beyond about 10°, because of the increase in spurious X-rays, but if there is a severe absorption problem, then decreasing t' by tilting the specimen toward the detector is a sensible first step toward minimizing the problem.

In some AEMs it is necessary to tilt the specimen toward the detector before any X-rays can be detected. Matters get even more complicated if your detector axis is not orthogonal to the tilt axis. Such a design is very poor from an analytical standpoint but, even under these conditions, the geometry is relatively straightforward and Zaluzec *et al.* (1981) have listed all the necessary equations.

| Remember, it is rare that you'll know the thickness of your specimen as well as you would like. |

So far we've assumed that the specimen is parallel-sided, but this is uncommon. Most thin-foil preparation methods result in wedge-shaped foils, and under these circumstances the detector must always be "looking" toward the thin edge of the specimen so that the X-ray path length is minimized, as we already mentioned in Figure 33.3. The only way to ascertain if this is a problem is to measure the thickness at each analysis point. Because this is such a tedious exercise, a method has been developed to correct for absorption without measuring t, as we discuss in the next section.

Because the sample density, ρ, and the values of μ/ρ vary with the composition of the specimen (see equations 35.22 and 35.23), the complete absorption correction procedure is an iterative process. The first step is to use the Cliff–Lorimer equation without any absorption correction and thus produce values for C_A and C_B. From these values, you perform a first iteration calculation of μ/ρ and ρ, and generate modified values of C_A and C_B, and so on. Usually, the calculation converges after two or three iterations.

In summary, there is substantial room for error in determining the various terms to insert into the ACF. For example, the ACF for k_{NiAl} in Ni_3Al, which is a strongly absorbing system, varies from ~5.5% to ~12% when the specimen doubles in thickness from 40 nm to 80 nm. This change is still quite small and within the limits of all but the most accurate microanalyses. In FeNi, which is a weakly absorbing system, a similar change in thickness would change the ACF for k_{FeNi} from ~0.6% to ~1.3%, which is negligible.

| It should always be remembered that large errors will only occur in strongly absorbing systems and/or very thick specimens. |

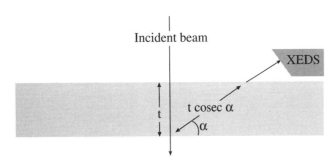

Figure 35.9. Relationship between the specimen thickness t and the absorption path length $t\,\mathrm{cosec}\,\alpha$ for a take-off angle α.

35.6. EXTRAPOLATION TECHNIQUES FOR ABSORPTION CORRECTION

A different approach to the absorption problem has been developed by Horita *et al.* (1987) and Van Cappellen (1990) which neatly avoids the problems of measuring the thickness at each analysis point, but does require that you measure the beam current. This is the way you should proceed with the absorption correction if it is at all possible.

You still need to know μ/ρ, ρ, and α, but not t. This approach uses a simplified correction factor

$$ACF' \simeq e^{-\left(\left.\frac{\mu}{\rho}\right|^B_{Spec} - \left.\frac{\mu}{\rho}\right|^A_{Spec}\right)\frac{\rho t}{2}\csc\alpha} \qquad [35.32]$$

which assumes that all X-rays are generated at $t/2$, ignoring $\varphi(\rho t)$ effects, and requires that the X-rays from one of the two elements are not absorbed. Applying the ACF' to the measured intensity ratios, we can show that, if we measure k_{AB} over a range of thicknesses, we can extrapolate to $(k_{AB})_0$ at $t = 0$ to give

$$\log_{10}(k_{AB}) = \log_{10}(k_{AB})_0 + \frac{\Delta_{AB}}{\varphi}I_x \qquad [35.33]$$

where Δ_{AB} is related to the difference in μ/ρ for X-rays from elements A and B:

$$\Delta_{AB} = 0.217\left(\left.\frac{\mu}{\rho}\right]^A_{Spec} - \left.\frac{\mu}{\rho}\right]^B_{Spec}\right)\rho\csc(\alpha) \qquad [35.34]$$

and

$$\varphi = C_A\left(\frac{Q\,\omega\,a}{A}\right)_A i_A \qquad [35.35]$$

for element A, where all the terms are described in equation 35.16, except for the electron probe current i_A, which is assumed constant.

So to apply this method, you need to keep the beam current and X-ray acquisition time constant, and the specimen must contain one X-ray that shows negligible absorption, as shown in Figure 35.10.

The method can be extended to the microanalysis of unknown specimens by using the extrapolation method to determine the absorption-free intensity ratio at zero thickness, and using this ratio in combination with the k_{AB} factor at zero thickness to give a value of C_A/C_B. This k factor is then applied to the calculation of the composition using the intensity ratio measured at the same thickness.

Obviously, it is very time-consuming to do the full absorption correction as accurately as we would like, because of all the uncertainties. You need an absorption correction if you are dealing with light elements, with K lines <1 keV. Under these circumstances, the extrapolation technique of Horita *et al.* is the best approach; a detailed example of the determination of light-element k factors using this method has been given by Westwood *et al.* (1992). Further refinements of Horita's method have been proposed by Eibl (1993). In the specific case of ionic compounds in which electroneutrality must be maintained (i.e., the sum of all anions and cations, times their valence states, must balance), it is even possible to devise an absorption correction with no estimate of t (Van Cappellen and Doukhan 1994).

35.7. THE FLUORESCENCE CORRECTION

X-ray absorption and fluorescence are intimately related because the primary cause of X-ray absorption is the fluorescence of another X-ray (such as the fluorescence of Si K_α X-rays in the XEDS detector which gives rise to the escape peak). You might think, therefore, that fluorescence corrections should be as widespread as absorption corrections. However, this is not the case for the following reasons. Strong absorption effects occur when there is a small amount of one element whose X-rays are being absorbed by the presence of a relatively large amount of another element. The absorption of Al K_α X-rays by Ni in Ni_3Al is a classic example. In this case, Ni X-rays are indeed fluoresced as a result of the absorption of Al K_α X-rays. However, there is a relatively small increase in the total number of Ni X-rays because Ni is the dominant element; the relative decrease in the Al K_α intensity is large because Al is the minor constituent. In this particular example there is a further reason why fluorescence of Ni X-rays is ignored; it is the Ni L_α X-rays which are fluoresced by the absorption of Al K_α X-rays. The Ni L X-rays are not the ones that we use for microanalysis anyhow, since the higher-energy Ni K X-rays are not absorbed or fluoresced.

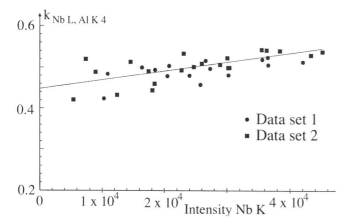

Figure 35.10. A plot of two independent sets of k-factor data for a Nb-Al alloy at 300 kV, showing the variation of the effective k factor with thickness as indicated by the Nb K_α X-ray intensity. The Al K_α X-rays which are absorbed give increasing effective k factors with thickness. X-rays for which absorption is insignificant would give a constant k factor with thickness.

> Fluorescence is usually a minor effect and often occurs for X-rays that are not of interest.

However, in the rare case that fluorescence occurs to a degree that limits the accuracy of microanalysis, the equation used for the fluorescence correction factor (FCF) is that developed by Nockolds *et al.* (1980); a detailed discussion is given by Anderson *et al.* (1995). Practical examples of the fluorescence correction are hard to come by and the classic case is Cr in stainless steels, where the Cr K_α line is fluoresced by the major peak, the Fe K_α line, giving rise to an increase in apparent Cr content as the foil gets thicker. You can also avoid the problem of thickness measurement for fluorescence corrections, just as for absorption, using a similar parameterless correction (Van Cappellen 1990).

35.8. ALCHEMI

We told you early on in this chapter to take your X-ray spectra away from strong diffraction conditions. This is because of the "Borrmann effect." As we saw back in Sections 13.8, 13.9, and 14.6, close to two-beam conditions the Bloch waves interact strongly with the crystal planes, and so X-ray emission is enhanced compared with kinematical conditions, as shown in Figure 35.11A. Now we can make use of this phenomenon to locate which atoms lie on which crystal planes. The technique has the delightful (and wholly inappropriate) acronym ALCHEMI, which is a selective abbreviation of the expression "Atom Location by CHanneling-Enhanced MIcroanalysis."

ALCHEMI is a quantitative technique for identifying the crystallographic sites, distribution, and types of substitutional impurities in many crystals. The technique was first developed for the TEM by Spence and Taftø (1983), who coined the acronym. The derivation of the quantitative expressions that we give below follows that paper. Channeling is widely used for atom site location in other analysis techniques (e.g., see Chu *et al.* 1978).

The way to do ALCHEMI experimentally is to acquire a spectrum under strong channeling conditions, such that the Bloch wave is interacting strongly with a particular systematic row of atoms. This channeling orientation should be chosen so that the planes interacting strongly with the electron beam also contain the candidate impurity atom sites, so you must have some *a priori* ideas about where substitutional atoms are most likely to sit. This technique is therefore particularly well suited to layer structures. When the Bloch wave is maximized on a particular plane of atoms, the X-ray intensity from the atoms in that

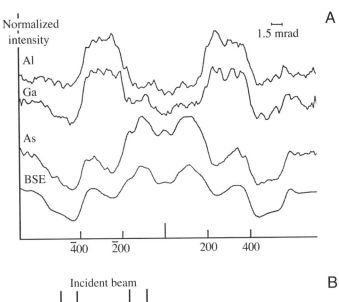

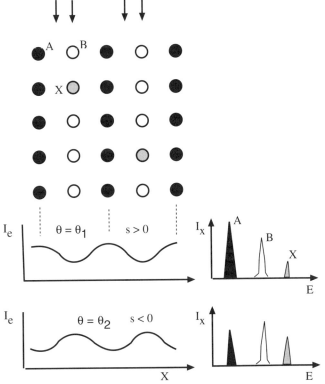

Figure 35.11. (A) The Borrmann effect: the variation in characteristic X-ray emission close to strong two-beam conditions as the beam is rocked across the 400 planes of GaAlAs. The X-rays from Al, which occupies Ga sites, follow the Ga X-ray emission while the As varies in an approximately complementary fashion. The backscattered electron signal (BSE) is inversely proportional to the amount of electron channeling, so the As signal is strongest where the channeling is weakest. (B) ALCHEMI allows the determination of the site occupancy of atom X in columns of atoms A and B. By tilting to **s** > 0 and **s** < 0, the Bloch waves interact strongly with row A and then row B, giving different characteristic intensities shown schematically in the spectra, from which the relative amounts of X in columns of A and B can be determined.

plane will be highest. Start by finding the orientations that give the most pronounced channeling effects for the atoms *A* and *B*, as shown schematically in Figure 35.11B. Usually a very small tilt is all that is necessary to get a different spectrum.

If you are looking at two elements, *A* and *B*, and a substitutional element *X*, follow this procedure:

■ Measure X-ray intensities from each element in orientations 1 and 2.
■ Then find a nonchanneling orientation (3) where electron intensity is uniform for both planes.

In this orientation we define the ratio *k* as

$$k = \frac{I_B}{I_A} \qquad [35.36]$$

where I_B is the number of X-ray counts from the element *B* in the nonchanneling orientation. For the two channeling orientations 1 and 2, we define two parameters β and γ such that

$$\beta = \frac{I_B^{(1)}}{k\, I_A^{(1)}} \qquad [35.37]$$

and

$$\gamma = \frac{I_B^{(2)}}{k\, I_A^{(2)}} \qquad [35.38]$$

Now assuming we know that the element *X* sits on specific sites, say it substitutes for atom *B*, then we define an intensity ratio term *R* such that

$$R = \frac{I_A^{(1)}\, I_X^{(?)}}{I_X^{(1)}\, I_A^{(2)}} \qquad [35.39]$$

Hence the fraction of atom *X* on *B* sites is given by

$$C_X = \frac{R-1}{R-1+\gamma-\beta R} \qquad [35.40]$$

Similar expressions can be generated for *X* atoms on *A* sites, but in fact the fraction of *X* atoms on *A* sites must be $1 - C_X$.

As you see, ALCHEMI can give a direct measure of the occupation of substitutional sites. However, the intensity differences in different orientations are often quite small and you need good X-ray statistics to draw sound conclusions. This makes it difficult to apply if high spatial resolution is also desired because, as we shall see in the next chapter, the conditions to give the best spatial resolution also give the worst counting statistics.

35.9. EXAMPLES; PROFILES AND MAPS

The best way to appreciate the value of quantitative analysis is to go and study some applications. In Figure 35.12, composition data from a complex three-component Ni-Cr-Mo high-temperature superalloy are plotted to reveal a section of the ternary phase diagram (Raghavan *et al.* 1984).

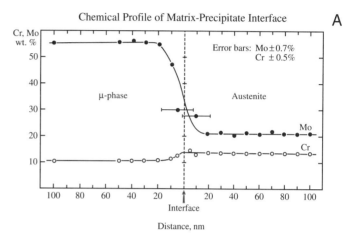

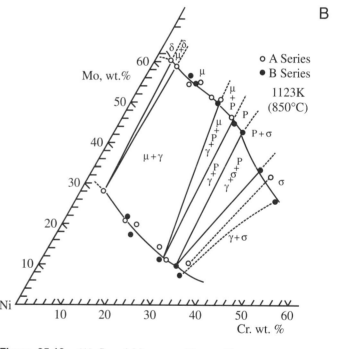

Figure 35.12. (A) Cr and Mo composition profiles across a two-phase μ–γ interface in a Ni-10Cr-30Mo alloy which has been aged 1000 hr at 1123 K. The profiles show composition changes that define tie lines in the ternary phase diagram. (B) A corner of the Ni-Cr-Mo ternary phase diagram determined by XEDS microanalysis of thin foils of heat-treated specimens containing up to three phases. The limits of the undesirable σ–phase regions are the important phase boundaries in this material.

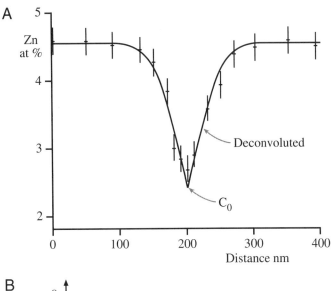

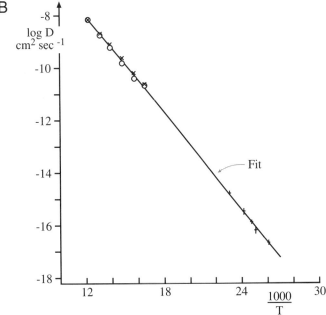

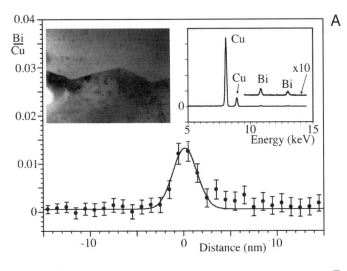

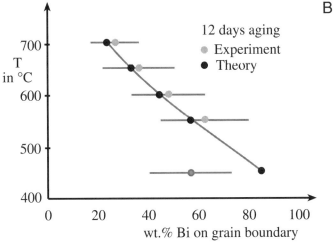

Figure 35.13. (A) Zn concentration profile across a grain boundary in an Al-4.5 at.% Zn alloy aged at 125°C to produce a solute-depleted region due to equilibrium grain boundary precipitation. Grube analysis of the profiles give a measure of D. The low aging temperatures produce such small profiles that AEM is the only way to measure them. (B) Arrhenius plot of the diffusivity of Zn in Al, as a function of temperature, derived from measurements of Zn composition profiles in (A). Extrapolation of high-temperature diffusion data match up well with the AEM results.

Figure 35.14. (A) XEDS measurements of the distribution of Bi segregated to a grain boundary in Cu. A faceted Cu GB, typical of those to which Bi is segregated, is shown in the upper-left inset and an XEDS spectrum from the GB region in the upper right reveals small Bi peaks. (B) Quantification of similar Bi spectra to that in (A) shows an inverse relationship with temperature, consistent with classical McLean adsorption isotherm predictions, shown as the fitted line.

Assuming interface equilibrium, sections such as the three-phase triangle can be measured from a single thin foil. In this study the undesirable σ–phase boundaries were sought, to avoid embrittlement of the alloy in service. In Figure 35.13, solute profiles measured across grain boundaries in Al-Zn (Nicholls and Jones 1983) permitted mea-surement of the diffusion coefficient of Zn in Al to much lower temperatures than previously attained with traditional EPMA methods. In Figure 35.14, the detection of Bi equilibrium segregation to grain boundaries in Cu is modeled by a simple McLean-type adsorption isotherm (Michael and Williams 1984). Previous studies of such Gibbsian segregation required *in situ* fracture of embrittled Cu inside Auger spectrometer systems.

We should also note that individual point analysis, or profiles across an interface, are not the only way to display X-ray data. It is possible to produce X-ray images or maps in which the intensity of the signal in the map is directly related to the X-ray intensity I_A. In a quantitative

A

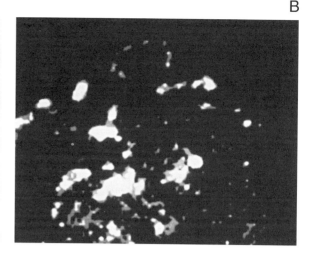

B

Figure 35.15.   (A) BF STEM image of PD catalyst particles on a carbon support film. (B) Qualitative Pd L_α X-ray image of the distribution of Pd.

map, the X-ray signal is proportional to the concentration C_A. While there are obvious advantages to comparing quantitative maps of elemental distributions with other TEM images, this process is limited by the relatively poor statistics of X-ray acquisition. Remember that good quantification requires ~10,000 counts for I_A. Even in an efficient AEM, this intensity may easily take one minute to acquire. At this acquisition rate, even a 56 × 56 pixel image will take 50 hours to gather, so it is impractical. We need to increase the efficiency of X-ray acquisition markedly, or we just have to make do with qualitative, noisy maps, as shown in Figure 35.15, even when using an FEG-AEM.

CHAPTER SUMMARY

Quantitative microanalysis of spectra from thin foils is straightforward in most cases, so long as you take care to determine the *k* factors with sufficient accuracy. The software to handle the more difficult problem of absorption is well known and commercially available. Perhaps the greatest difficulty remains the need to know the specimen thickness in order to compensate for X-ray absorption, and extrapolation techniques are invaluable in avoiding this. We can minimize absorption by making the thinnest possible specimens, but then the possibility arises that the number of X-ray counts may be so small that errors in the quantification are large. The use of FEG sources and improved TEM-EDS configurations to maximize the collection angle will help in this situation.

REFERENCES

General References

Goldstein, J.I. (1979) in *Introduction to Analytical Electron Microscopy* (Eds. J.J. Hren, J.I. Goldstein, and D.C. Joy), p. 83, Plenum Press, New York.

Goldstein, J.I., Williams, D.B., and Cliff, G. (1986) in *Principles of Analytical Electron Microscopy* (Eds. D.C. Joy, A.D. Romig, and J.I. Goldstein), p. 155, Plenum Press, New York.

Romig A.D. Jr. (1986) *Analytical Transmission Electron Microscopy,* Metals Handbook, 9th edition, **10**, p. 429, American Society for Metals, Metals Park, Ohio.

Williams, D.B. (1987) *Practical Analytical Electron Microscopy in Materials Science,* 2nd edition, Philips Electron Optics Publishing Group, Mahwah, New Jersey.

Williams D.B. and Goldstein J.I. (1991) in *Electron Probe Quantitation* (Eds. K.F.J. Heinrich and D.E. Newbury), p. 371, Plenum Press, New York.

Zaluzec, N.J. (1979) in *Introduction to Analytical Electron Microscopy* (Eds. J.J. Hren, J.I. Goldstein, and D.C. Joy), p. 121, Plenum Press, New York.

Specific References

Anderson, I.M., Bentley, J., and Carter, C.B. (1995) *J. Microsc.* **178**, 226.

Bambynek, W., Crasemann, B., Fink, R.W., Freund, H.U., Mark, H., Swift, C.D., Price, R.E., and Rao, P.V. (1972) *Rev. Mod. Phys.* **44**, 716.

Bender, B.A., Williams, D.B., and Notis, M.R. (1980) *J. Am. Ceram. Soc.* **63**, 149.

Castaing, R. (1951) Thesis, University of Paris, ONERA Publication, #55.

Chapman, J.N., Nicholson, W.A.P., and Crozier, P.A. (1984), *J. Microsc.* **136**, 179.

Chu, W.-K., Mayer, J.M., and Nicolet, M.-A. (1978) *Backscattering Spectrometry,* Academic Press, Orlando, Florida.

Cliff, G. and Lorimer, G.W. (1975) *J. Microsc.* **103**, 203.

Eibl, O. (1993) *Ultramicrosopy* **50**, 179.

Goldstein, J.I., Costley, J.L., Lorimer, G.W., and Reed, S.J.B. (1977) *SEM 1977,* **1** (Ed. O. Johari), p. 315, IITRI, Chicago, Illinois.

Graham, R.J. and Steeds, J.W. (1984) *J. Microsc.* **133**, 275.

Heinrich, K.F.J. (1986) in *Proc. ICXOM-11* (Eds. J. Brown and R. Packwood), p. 67, University of Western Ontario, Canada.

Heinrich, K.F.J. and Newbury, D.E., Eds. (1991) *Electron Microprobe Quantitation,* Plenum Press, New York.

Hillier, J. and Baker, R.F. (1944) *J. Appl. Phys.* **15**, 663.

Horita, Z., Sano, T., and Nemoto, M. (1987) *Ultramicroscopy* **21**, 271.

Kramers, H.A. (1923) *Phil. Mag.* **46**, 836.

Lorimer, G.W., Al-Salman, S.A., and Cliff, G. (1977) in *Developments in Electron Microscopy and Analysis* (Ed. D.L. Misell), p. 369, The Institute of Physics, Bristol and London.

McGill, R.H. and Hubbard, F.H. (1981) in *Quantitative Microanalysis with High Spatial Resolutions* (Eds. G.W. Lorimer, M.H. Jacobs, and P. Doig), p. 30, The Metals Society, London.

Michael, J.R. and Williams, D.B. (1984) *Met. Trans.* **15A**, 99.

Nicholls, A.W. and Jones, I.P. (1983) *J. Phys. Chem. Solids* **44**, 671.

Nockolds, C., Nasir, M.J., Cliff, G., and Lorimer, G.W. (1980) in *Electron Microscopy and Analysis-1979* (Ed. T. Mulvey), p. 417, The Institute of Physics, Bristol and London.

Owen, D.B. (1962) *Handbook of Statistical Tables,* Addison-Wesley, Reading, Massachusetts.

Powell, C.J. (1976) in *Use of Monte Carlo Calculations in Electron Probe Microanalysis and Scanning Electron Microscopy* (Eds. K.F.J. Heinrich, D.E. Newbury, and H. Yakowitz), p. 61, U.S. Department of Commerce/NBS, Washington, DC.

Raghavan, M., Mueller, R.R., Vaughn, G.A., and Floreen, S. (1984) *Met. Trans.* **15A**, 783.

Schreiber, T.P. and Wims, A.M. (1981) *Ultramicroscopy* **6**, 323.

Schreiber, T.P. and Wims, A.M. (1982) *X-ray Spectrometry* **11**, 42.

Sheridan, P.J. (1989) *J. Electr. Microsc. Tech.* **11**, 41.

Spence, J.C.H. and Taftø, J. (1983) *J. Microsc.* **130**, 147

Sprys, J.W. and Short, M.A. (1976) *Proc. 34th EMSA Meeting* (Ed. G.W. Bailey), Claitors, Baton Rouge, Louisiana.

Van Cappellen, E. (1990) *Microsc. Microanal. Microstruct.* **1**, 1.

Van Cappellen, E. and Doukhan, J.C. (1994) *Ultramicroscopy* **53**, 343.

Westwood, A.D., Michael, J.R., and Notis, M.R. (1992) *J. Microsc.* **167**, 287.

Williams, D.B. and Goldstein, J.I. (1991) in *Electron Probe Quantitation* (Eds. K.F.J. Heinrich and D.E. Newbury), p. 371, Plenum Press, New York.

Williams, D.B., Newbury, D.E., Goldstein, J.I., and Fiori, C.E. (1984) *J. Microsc.* **136**, 209.

Wood, J.E., Williams, D.B., and Goldstein, J.I. (1981) in *Quantitative Microanalysis with High Spatial Resolutions* (Eds. G.W. Lorimer, M.H. Jacobs, and P. Doig), p. 24, The Metals Society, London.

Wood, J.E., Williams, D.B., and Goldstein, J.I. (1984) *J. Microsc.* **133**, 255.

Zaluzec, N.J., Maher, D.M., and Mochel, P.E. (1981) in *Analytical Electron Microscopy-1981* (Ed. R.II. Geiss), p. 25, San Francisco Press, San Francisco, California.

Spatial Resolution and Minimum Detectability

36

CHAPTER PREVIEW

Often, when you do X-ray microanalysis of thin foils, you are seeking information that is close to the limits of spatial resolution. Before you carry out any such microanalysis you need to understand the various controlling factors, which we explain in this chapter. Minimizing your specimen thickness is perhaps the most critical aspect of obtaining the best spatial resolution, so we summarize the various ways you can measure your foil thickness at the analysis point.

A consequence of going to higher spatial resolution is that the X-ray signal comes from a much smaller volume of the specimen. A smaller signal means that we find it very difficult to detect the presence of trace constituents in thin foils. Consequently, the minimum mass fraction (MMF) in AEM is not very small compared with other analytical instruments which have poorer spatial resolution. This trade-off is true for any microanalysis technique, and so it is only sensible to discuss the ideas of spatial resolution in conjunction with analytical detectability limits. We'll make this connection in the latter part of the chapter. Despite the relatively poor MMF we can detect the presence of just a few atoms of one particular element if the analyzed volume is small enough, and so the AEM actually exhibits excellent minimum detectable mass (MDM) characteristics.

Spatial Resolution and Minimum Detectability

36

36.1. WHY IS SPATIAL RESOLUTION IMPORTANT?

As we described in the introduction to Chapter 35, perhaps the major driving force for the development of X-ray microanalysis in the AEM was the improvement in spatial resolution compared with the EPMA. This improvement arises for two reasons:

- We use thin specimens, so less electron scatter occurs as the beam traverses the specimen.
- The higher electron energy (>100–400 keV in the AEM compared with 5–30 keV in the EPMA) further reduces scatter.

The latter effect occurs because the mean-free path for both elastic and inelastic collisions increases with the electron energy. The net result is that *increasing* the accelerating voltage when using thin specimens *decreases* the total beam–specimen interaction volume, thus giving a more localized X-ray signal source and a higher spatial resolution, as you can see in the Monte Carlo simulations in Figure 36.1A. Conversely, with bulk samples, *increasing* the voltage *increases* the interaction volume, and spatial resolution rarely improves below ~0.5 μm, as shown in Figure 36.1B. So no one is very interested in the theory of spatial resolution of microanalysis for bulk samples and little effort, beyond lowering the kV, is routinely made to optimize this parameter in practice. By contrast, much theoretical and experimental work has been carried out to both define and measure the spatial resolution of XEDS in the AEM, and we'll introduce some of the major ideas here.

36.2. DEFINITION OF SPATIAL RESOLUTION

We can define the spatial resolution of X-ray microanalysis as the smallest distance (R) between two volumes from which independent X-ray microanalyses can be obtained. The definition of R has evolved as AEMs have improved and smaller analysis volumes have become attainable. It has long been recognized that the analysis volume, and hence R, is governed by the beam–specimen interaction volume, since the XEDS can detect X-rays generated anywhere within that volume. The interaction volume is a function of the incident beam diameter (d) and the beam spreading (b) caused by elastic scatter of the beam within the specimen. Therefore, the measured spatial resolution is a function of your specimen, and this has made it difficult to define a generally accepted measure of R. Let's look first at d and b and how we define them.

We've already discussed how to define and measure d in TEMs and STEMs way back in Chapter 5, so you need only remind yourself that

> The beam diameter, d, is customarily defined as the FWTM of the Gaussian electron intensity. You can measure d directly from the TEM viewing screen, or indirectly by traversing the beam across a sharp edge and looking at the intensity change on the STEM screen.

This definition takes account of only 90% of the electrons entering the specimen, so it is an approximation. Remember that the intensity distribution in the incident beam is Gaussian only if you are careful in your choice and alignment of the C2 aperture. It is a little more diffi-

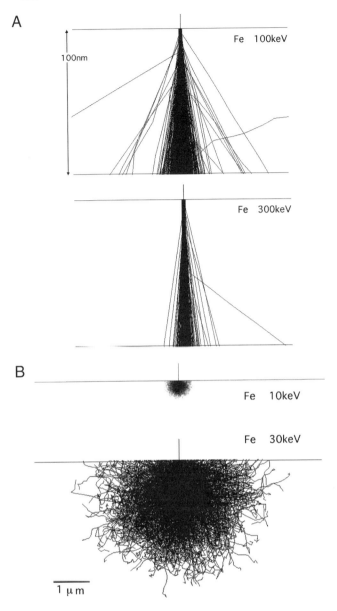

Figure 36.1. (A) Monte Carlo simulations of 10^3 electron trajectories through a 100-nm thin foil of Fe at 100 kV and 300 kV. Note the *improved* spatial resolution at higher kV. (B) Conversely, in a bulk sample the interaction volume at 30 kV is significantly more than at 10 kV, giving *poorer* X-ray spatial resolution at higher kV.

cult to define and measure b, so this needs more explanation.

36.3. BEAM SPREADING

The amount that the beam spreads on its way through the specimen (b) has been the subject of much theoretical and experimental work. While results and theories differ in minor aspects, there is a general consensus that b is governed by the beam energy (E_0), foil thickness (t), and density (ρ). It turns out that the most simple theory for b usually gives a good approximation under most microanalysis conditions. This theory, sometimes called the "single-scattering" model because it assumes that each electron only undergoes one elastic scattering event as it traverses the specimen, was first given in the seminal paper by Goldstein *et al.* (1977), and refined by Reed (1982). This single-scattering model states

$$b = 7.21 \times 10^5 \frac{Z}{E_0} \left(\frac{\rho}{A} \right)^{\frac{1}{2}} t^{\frac{3}{2}} \qquad [36.1]$$

This well-known expression is *not* in SI units because b and t are in given in cm, ρ is in g/cm^3, and E_0 is in eV.

This definition again comprises 90% of the electrons emerging from the specimen, so it is consistent with our definition of d.

There is some question as to whether this expression adequately describes the behavior of b for either very thin or very thick foils, but it has generally survived the test of time and its strength remains in its simplicity.

> We recommend that you keep this equation stored in your calculator or in the AEM computer system, so you can quickly estimate the expected beam spreading in your planned experiment.

You should, of course, estimate b *prior* to spending an inordinate amount of time trying to do an experiment which is impossible for lack of spatial resolution.

When we can't apply equation 36.1 (for example, if the specimen geometry or microstructure is complex) the best alternative is the Monte Carlo computer simulation (see Figure 36.1), which we introduced in Section 2.6 as a way of modeling electron scatter. Remember that the Monte Carlo technique uses a random number generator (hence the name) to simulate elastic and inelastic electron–specimen interactions and generate a feasible set of electron paths through a defined specimen. A relatively small number of paths (typically 10^3–10^6) can give a very good measure of the behavior of the very large number of electrons in a typical beam (remember, a 1-nA probe current implies ~10^{10} electrons/second entering the specimen). A full description is beyond the scope of this text and complete books exist on the topic (Heinrich *et al.* 1975, Joy 1995). After simulating several thousand paths, an approximate value of b can be obtained by asking the computer to calculate the diameter of a disk at the exit surface of the specimen that contains 90% of the emerging electrons. This definition of b is consistent with that described at the start and is the dimension of b given by equation 36.1. In Joy's book, you'll find a code listing for a Monte Carlo simulation program which can be run on a PC. These simulations are now extremely rapid, and in a few minutes on a PC they can provide much of the information you need to estimate the beam spreading in heterogeneous microstructures that are not amenable to simple modeling with the single-scattering approach. Parallel supercomputers have even been used to simulate millions of trajectories in more complex specimens (Michael *et al.* 1993).

While beam spreading is the main aspect of spatial resolution theories, we mustn't forget that what we really want to know is the beam–specimen interaction volume, which corresponds to the X-ray source size. Of course this is closely related to the electron distribution, but we can only relate the two directly using Monte Carlo simulations. In these simulations you can easily get the computer to calculate the distribution of X-ray photons generated throughout the specimen, and factor it into any calculations of the composition of the analyzed volume. Monte Carlo simulations are useful for estimating the X-ray spatial resolution because they:

■ Incorporate the effects of different kVs and beam diameters.
■ Handle difficult specimen geometries and multiphase specimens.
■ Automatically calculate the effect of the depth distribution of X-ray production $\varphi(\rho t)$ on the X-ray source size.
■ Display the X-ray distribution generated anywhere in your specimen as a function of all its parameters, ρ, Z, A, and t. This tells you the relative contributions to your XEDS spectrum from different parts of the microstructure.

In addition to the theories of beam spreading that we've discussed, there are several more in the literature. A common feature of these theories is that they all predict a linear relationship between b and $t^{3/2}$ and an inverse relationship between b and E_0. If you're interested in the details of the various theories you'll find a discussion in Goldstein *et al.* (1986).

36.4. THE SPATIAL RESOLUTION EQUATION

Now that we've defined d and b, all we have to do is combine them to come up with a definition of R. Reed (1982) argued that if the incident beam was Gaussian, and if the

beam emerging from the specimen retains a Gaussian intensity distribution, then b and d should be added in quadrature to give a value for R

$$R = \left(b^2 + d^2\right)^{\frac{1}{2}} \qquad [36.2]$$

This equation remained the standard definition of R for almost a decade, despite the fact that no set of experiments ever investigated the effects of all the variables affecting b in equation 36.1. About the same time as this definition of R was proposed, Gaussian beam-broadening models were introduced which were based on equation 36.1 but permitted convolution of the Gaussian descriptions of d and b to come up with a definition of R. Based on the Gaussian model and experimental measurements, Michael *et al.* (1990) proposed that the definition of R be modified so as not to present the worst case (given by the exit beam diameter) but to define R midway through the foil, as shown in Figure 36.2

$$R = \frac{d + R_{max}}{2} \qquad [36.3]$$

where R_{max} is given by equation 36.2.

> This equation is the formal definition of the X-ray spatial resolution.

Like all definitions of spatial resolution, there is no fundamental justification for the choice of various factors

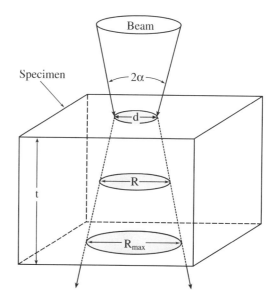

Figure 36.2. Definition of spatial resolution: schematic diagram of how the combination of incident beam size d and beam spreading through the foil combine to define the spatial resolution R of X-ray microanalysis in a thin foil.

such as the FWTM diameter, and the selection of the mid-plane of the foil at which to define R. Similarly, this approach ignores any contribution of electron diffraction in crystalline specimens, beam tailing beyond the 90% limit, and the effects of fast secondary electrons, which, in some circumstances, can be important. Nevertheless, the definition has been shown to be consistent with experimental results and sophisticated Monte Carlo simulations (Williams *et al.* 1992). Finally, this definition retains the advantage of the original single-scattering model, i.e., it has a simple form, easily amenable to calculation.

36.5. MEASUREMENT OF SPATIAL RESOLUTION

Any theory of spatial resolution must be tested against practical measurements in the TEM if it is to be relevant. Experimental measurements of the spatial resolution appeared slightly before the first theoretical treatments. Composition profiles measured across atomically sharp interphase interfaces were first presented by Lorimer *et al.* (1976). Since then, several other kinds of specimens have been proposed, such as spherical particles in a foil of known thickness, artificial specimens of Au lines deposited on a Si foil, grain boundary films, and quantum well structures, among others.

> We believe the first method, using interphase interfaces, retains its validity since there are fewer unknowns than for the other specimens.

If thermodynamic equilibrium exists either side of the interphase interface, the solute content of each phase is well defined. Also, interphase interfaces are common to many engineering materials, as shown back in Figure 35.12.

In order to compare experimental and calculated measurements of R, you have to understand how we relate the measured composition profile across the interface to the actual discrete profile shape, shown schematically in Figure 36.3. We do this by deconvolution of the beam shape from the measured profile. The finite beam size, d, and the effect of b degrade the sharp profile to a width L, which is related to R by the following equation:

$$R = 1.414L \qquad [36.4]$$

Assuming this relationship holds, we just measure the distance L between the 2% and 98% points on the profile, as shown in Figure 36.3. This spread contains 90% of the beam electrons, consistent with our assumption of a 90%

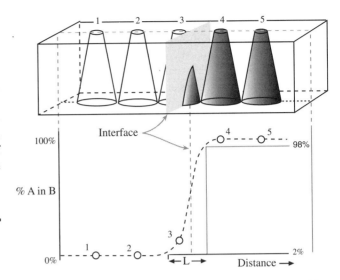

Figure 36.3. Schematic diagram showing the measured composition profile obtained across a planar interface at which an atomically discrete composition change occurs. The spatial resolution can be related to the extent (L) of the measured profile between the 2% and 98% points.

(FWTM) incident beam diameter. In practice, you will find it difficult to measure the 2% and 98% points because of the errors in the experimental data. So you should measure the distance from the 10% to the 90% points on your profile, corresponding to the beam spread containing 50% of the electrons (FWHM), then multiply this distance by 1.8 to give the FWTM.

> Note that this definition of R, like the definitions of b and d that we have used, is arbitrary.

Nevertheless, it is easy to remember, relatively easy to measure, consistent with the definitions of b and d, and, most importantly, gives a number that is close to the experimentally measured degradation of the discrete composition change introduced by the beam and the specimen.

Typical measurements of the spatial resolution in two different AEMs are shown in Figures 36.4 and 36.5. Two composition profiles are shown. Each was taken from an Fe-Ni-Cr foil aged to give large Cr composition changes between Cr-rich α-Cr precipitates and the Cr-poor matrix. The specimen was aged sufficiently that the precipitate and matrix are in thermodynamic equilibrium, so that a discrete (atomic level) composition change occurs at the interface. The smooth profiles that appear in the figure are then due to the effects of b and d. In Figure 36.4, the data were obtained from an FEG AEM in which the accelerating voltage was kept constant at 100 kV and the specimen thickness

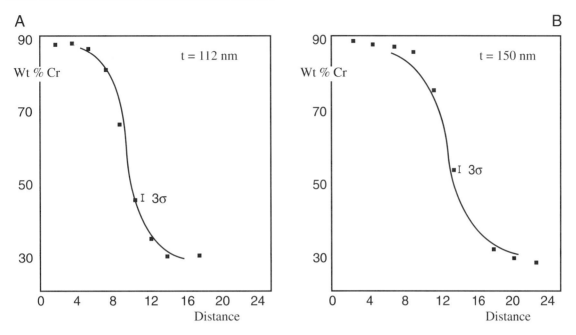

Figure 36.4. Measured Cr composition profiles across the same interface in an Fe-Ni-Cr alloy at two different thicknesses, (A) 112 nm and (B) 150 nm. The solid lines are the fits to the experimental data obtained using a Gaussian convolution model. The profiles were obtained in an FEG AEM at 100 kV and show that the $t^{3/2}$ relationship, assumed in the Gaussian model, applies over the range of thicknesses studied.

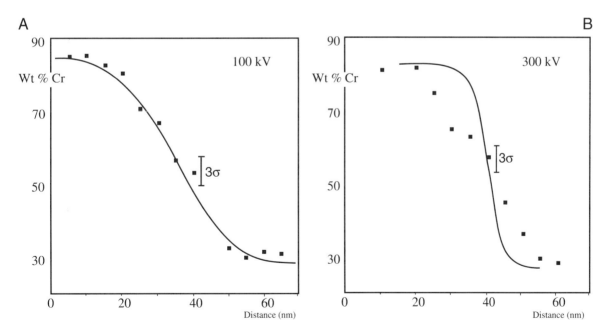

Figure 36.5. Composition profiles from an Fe-Ni-Cr foil, 112 nm thick, obtained with a thermionic source AEM at (A) 100 kV and (B) 300 kV. The solid lines are the fits obtained with a Gaussian convolution model and demonstrate poor spatial resolution, dominated by the large beam size. The bad fit at 300 kV indicates possible specimen drift.

was varied from 112 nm to 150 nm. In Figure 36.5, the data were obtained on an intermediate voltage AEM in which the foil thickness was a constant 120 nm but the voltage was varied between 100 kV and 300 kV. The line drawn through the experimental measurements in each case is derived from the Gaussian model.

It is obvious from equation 36.2 that if we want to improve spatial resolution, then both d and b must be minimized. But if we minimize d, we reduce the input beam current and, for thermionic sources, if $d < 10$ nm, count rates are unacceptably low. However, with an FEG, sufficient current (~1 nA) can be generated in a 1-nm beam to permit quantitative analysis. Comparison of the 100-kV data from the thermionic source instrument (Figure 36.5A) and the FEG source (Figure 36.4A) shows the improvement gained by the use of very small beams in the FEG instrument.

So if you have a thermionic source AEM:

- Your specimen has to be thick enough that sufficient counts are generated for quantification and the net result may be that b is the main contributor to R.
- Alternatively, you may have to increase the beam size such that d dominates rather than b, as in Figure 36.5A (beam diameter 56 nm).

Such a large beam was needed in that example in order to generate sufficient beam current to get a reasonable X-ray count rate at 100 kV. This is one reason why there's been a lot of effort put into developing 300–400 kV AEMs and, more recently, 200–300 kV FEG AEMs.

There are some practical factors which can also limit your experimental spatial resolution:

- Specimen drift and carbon contamination are real problems with side-entry goniometer stages.
- Drift is often exacerbated by the liquid-N_2 cooling required to minimize carbon contamination.
- Changing the kV in intermediate voltage instruments subjects the objective lens cooling coil to large changes in thermal load, which causes drift.

Improvements in image analysis software mean that on-line drift correction is now available. If you're planning to carry out microanalysis at the highest spatial resolution where you're obliged to count for long times to accumulate adequate X-ray intensity, then such software is indispensable. Perhaps one unfortunate side effect of higher voltages is that analysis can be performed in thicker areas than at 100 kV and spatial resolution degrades.

In summary, the spatial resolution R is a function of both the beam size and the beam spreading. You can get a good estimate of R from equation 36.3. The theories all indicate a $t^{3/2}$ dependence of the beam spreading, so thin specimens are essential for the best resolution. FEG sources give sufficient beam current to generate reasonable counts even from very thin specimens and invariably give the best spatial resolution.

36.6. THICKNESS MEASUREMENT

Given the $t^{3/2}$ dependence of the beam spreading, you can see the importance of knowing t when estimating the spatial resolution. You already know that t is also an essential parameter in correcting for the absorption and fluorescence of characteristic X-rays, as we saw in the previous chapter. Furthermore, you should remember that knowledge of t is important in high-resolution phase-contrast imaging and CBED. You'll see in Chapter 39 that in EELS, minimizing t is again critical to obtaining the best results. In almost all TEM techniques your specimen has to be as thin as possible to get the best results; CBED studies are a notable exception to this generalization.

So let's take the opportunity here to summarize the methods available for measuring thickness. The methods are many and varied, and a full discussion of the most important techniques will be found in other parts of this book. The first point to remember is that the thickness we are interested in is t, the thickness through which the beam penetrates. This value depends both on the tilt of the specimen γ, and the true thickness at zero tilt, t_0. As shown in Figure 36.6, for a parallel-sided foil

$$t = \frac{t_0}{\cos \gamma} \qquad [36.5]$$

If your specimen is wedge-shaped, then t and t_0 will vary in an arbitrary fashion depending on the foil shape.

36.6.A. TEM Methods

In the TEM you can get always make an estimate of your specimen thickness if it is wedge-shaped (and crystalline). By tilting to two-beam conditions for strong dynamical diffraction, the BF and DF images both show thickness fringes, as we saw in Chapter 23. These fringes occur at regions of constant thickness. The intensity in the BF image falls to zero at a thickness of $0.5\xi_g$ at $\mathbf{s} = 0$. Therefore, to determine t all you have to do is look at the BF image and count the number (n) of dark fringes from the edge of the specimen to the analysis region. At that point $t = (n - 0.5)\xi_g$, assuming that the thinnest part at the

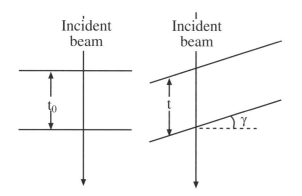

Figure 36.6. The difference between the specimen thickness, t_0, and the distance traveled by the beam, t, that determines beam spreading in a specimen tilted through an angle γ.

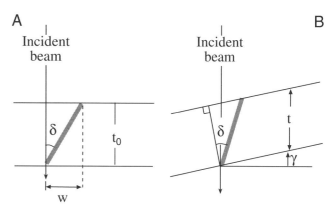

Figure 36.7. The parameters required to measure foil thickness t from a planar defect (projected width w), inclined to the incident beam by angle δ; comparison of (A) an untilted specimen normal to the beam and (B) a specimen tilted through an angle γ.

edge is $< 0.5\xi_g$ thick. (Be very careful with this assumption.) Remember that the value of ξ_g varies with diffracting conditions and so the **g**-vector has to be specified. You can calculate ξ_g from the expression

$$\xi_g = \frac{\pi \, \Omega \, \cos\theta}{\lambda \, f(\theta)} \qquad [36.6]$$

where Ω is the volume of the unit cell, λ is the electron wavelength, and $f(\theta)$ is the atomic scattering amplitude. Remember also that if you're not exactly at $\mathbf{s} = 0$, then the effective extinction distance ξ_{eff} must be used.

A related method relies on the presence of an inclined planar defect adjacent to the analysis region. The projected image of the defect, again under two-beam conditions, will exhibit fringes, which can be used to estimate the local thickness, or the projected width, w, of the defect image using the expression

$$t_0 = w \cot \delta \qquad [36.7]$$

as shown in Figure 36.7, in which δ is the angle between the beam and the plane of the defect. Again, you have to compensate geometrically to measure t rather than t_0 if the foil isn't normal to the beam, and then

$$t = w(\cot \delta - \tan \gamma) \qquad [36.8]$$

Of course both of these methods are inapplicable to noncrystalline materials, and it is not always possible to find a suitable inclined defect next to the analysis region. Furthermore, two-beam conditions are not recommended for microanalysis because of the dangers of anomalous X-ray emission (see Section 35.8). More insidious is the fact that oxidation, during or after specimen preparation, means that your crystalline specimen may be coated with an amor-

phous layer, which will not be measured by these diffraction-contrast techniques.

Another method related to the TEM image contrast involves measurement of the relative transmission of electrons. The intensity on the TEM screen decreases with increasing thickness, all other things being equal. Make all the intensity measurements on your specimen under the same diffraction conditions and incident beam current but with no objective aperture. By calibrating the intensity falling on the screen with a Faraday cup, you can get a crude measure of relative thickness, which can be converted into an absolute measure of t if some absolute method is used for calibration. The only advantage of this approach is that it is applicable to all materials, both amorphous and crystalline, but it is tedious and not very accurate.

Finally, an old method for thickness determination was to deposit small latex spheres on either side of your specimen and measure the thickness by noting parallax shifts between balls on the top and bottom sides as you tilt. This is not recommended, because the latex solution will contribute to specimen contamination and there are alternative and better methods. However, there are cases where you'll have particles or other markers already present on both surfaces, so you might use these for the parallax method.

36.6.B. Contamination-Spot Separation Method

This method, unique to a probe-forming (S)TEM, relies on the propensity of such instruments to generate carbon contamination on both the top and bottom surfaces of the specimen at the point of analysis. If you tilt your specimen by a

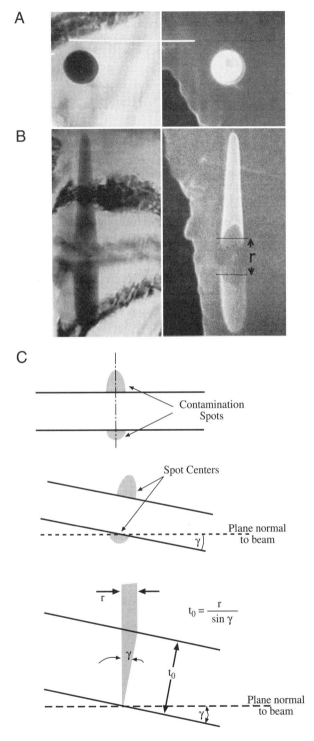

Figure 36.8. The contamination-spot separation method for thickness determination; (A) the contamination is deposited on both surfaces of the specimen and the separation (r) is only visible in (B) when the specimen is tilted sufficiently. The images show the contamination at zero tilt and high tilt angle γ. The two images on the left are obtained in STEM BF and on the right in SE mode. SE mode gives the best image contrast. (C) Geometry required to determine thickness t_0 from the projected spacing r of the contamination spots.

large enough angle γ, you can see discrete contamination spots (Figure 36.8). Their separation r, at a screen magnification M, is related to t_0 by the following expression

$$t_0 = \frac{r}{M \sin \gamma} \qquad [36.9]$$

If the specimen itself is tilted by an angle ε when the contamination is deposited, then

$$t_0 = \frac{r \cos \varepsilon}{M \sin \gamma} \qquad [36.10]$$

Although this method is straightforward, it relies on highly undesirable contamination, which obscures the very area you're looking at! Contamination degrades the spatial resolution and increases the X-ray absorption. In fact, we spend a lot of time and effort trying to minimize contamination, so it would be perverse to propose it as a useful way of determining t. Having said that, and despite ample evidence that the spot separation method overestimates the thickness by as much as 100%, it is often used because it is quick (and dirty!); it measures t exactly at the analysis point and the shape of the spots can indicate if the beam or the specimen has drifted during microanalysis. So if you can't avoid it, use it with caution.

36.6.C. X-ray Spectrometry Methods

Your X-ray spectrum intensity is a measure of the specimen thickness. Indeed, the standard method of quantification in biological materials uses the bremsstrahlung intensity as a measure of the mass thickness of the specimen, although this isn't used by materials scientists. If an element has two characteristic X-ray lines visible in the spectrum, e.g., the L and K lines, then the relative intensity of these lines will change as the specimen thickness increases because the lower-energy line will be more strongly absorbed. Knowing the necessary absorption parameters, such as μ/ρ, it is then possible to deduce ρt by an iterative process, which is essentially the same as the absorption correction discussed in Section 35.5. If ρ is known, then t can be determined. A similar method involves recording spectra containing an X-ray line that is strongly absorbed at two different tilts, or using two detectors with different take-off angles. In such cases, the two spectra will show different characteristic peak intensities because of the different absorption path lengths in each case. Again, an iterative process is required to extract t.

Porter and Westengen (1981) proposed such a method using standards, which still relies on an iterative procedure based on the absorption correction. In this method, the mass thickness is determined from the following equation

$$\rho t = \frac{\cos \gamma I_0(A)}{e_A \, m_A} \qquad [36.11]$$

where γ is the tilt angle, m_A is the mass fraction of element A in the specimen, e_A is the number of counts from element A detected per unit incident electron, per unit mass thickness of A in the absence of absorption, and $I_0(A)$ is the absorption-corrected intensity of A from the specimen. The calibration constant e_A is obtained from a standard of known mass thickness, such as a pure element foil. Since $I_0(A)$ is unknown, it is obtained from the absorption correction equation given by

$$I_0(A) = - \frac{\ln \left[1 - u_A \, I(A) \right]}{u_A} \qquad [36.12]$$

where $I(A)$ is the observed intensity of X-rays from element A and u_A is given by

$$u_A = \frac{\left. \frac{\mu}{\rho} \right|_{Spec}^{A} \cosec \alpha \cos \theta}{e_A \, m_A} \qquad [36.13]$$

You can write a similar equation for any element in your specimen. Thus, once your standard has been calibrated, the mass thickness can easily be calculated. However, as in all the absorption correction methods, an iterative process is required. If the foils are bent, or change thickness rapidly, then these methods all become very difficult to carry out and rather inaccurate. Furthermore, if e_A is to be a reliable calibration factor, the electron beam current must be stable and easily measurable; this isn't usually the case in the AEM.

Another closely-related method, described in Section 35.3 when we were discussing absorption, is the extrapolation technique in which the X-ray intensity is related directly to the mass thickness. So if you know the specimen composition you can obtain a value for t (Horita *et al.* 1989).

36.6.D. Electron Energy-Loss Spectrometry Methods

Thickness information is present in the electron energy-loss spectrum, since the intensity of inelastically scattered electrons increases with your specimen thickness. In essence, you have to measure the intensity under the zero-loss peak (I_0) and ratio this to the total intensity in the spectrum (I_T). The relative intensities are governed by the mean free path (λ) for energy loss. A parameterization formula for λ is discussed in detail in Section 39.5.

We can apply the EELS method to any specimen, amorphous or crystalline.

EELS is applicable over a wide range of thicknesses, and with parallel-collection spectrometers it is so rapid that you can even produce thickness "maps" of thin foils. So this approach is highly recommended.

36.6.E. Convergent-Beam Diffraction Method

The CBED pattern, which is visible on the TEM screen when a convergent beam is focused on the specimen, can also be used to determine the thickness of crystalline specimens. In Section 21.1 we described the procedure to extract the thickness from the K-M fringe pattern obtained under two-beam conditions. The CBED pattern must come from a region thicker than $1\xi_g$ or else fringes will not be visible. Also, the region of the foil should be relatively flat and undistorted. We can envisage on-line thickness determination by digitizing the CBED pattern, scanning it across the STEM BF detector, and measuring the fringe spacing from the Y-modulation output on the STEM CRT. For clean crystalline specimens, this is *the* way to determine t.

In summary, there are many ways you can determine t, but no one method is convenient, accurate, and universally applicable. The various methods also measure different thicknesses, e.g., the crystalline thickness, neglecting surface films, or the thickness including surface films, or the mass thickness. The EELS, CBED, and X-ray absorption methods all have the possibility of widespread on-line use, and we recommend these methods, in order of preference. Detailed reviews of the methods of determining t have also been given by Berriman *et al.* (1984) and Scott and Love (1987).

36.7. MINIMUM DETECTABILITY

Minimum detectability is a measure of the smallest amount of a particular element that can be detected with a defined statistical certainty. Minimum detectability and spatial resolution are intimately related.

It is a feature of any microanalysis technique that an improvement in spatial resolution is balanced by a worsening of the detectability limit (all other factors being equal).

At higher spatial resolution the analyzed volume is smaller, and therefore the signal intensity is reduced. This reduction in signal intensity means that the acquired spectrum will be noisier and small peaks from trace elements will be less detectable. Accordingly, in the AEM, the price that you pay for improved spatial resolution is a relatively poor minimum detectability. By way of comparison, Figure 36.9 compares the size of the analyzed volume in an EPMA, a TEM/STEM with a thermionic source, and a dedicated STEM with an FEG. The enormous reduction in the beam–specimen interaction volume explains the small signal levels that we obtain in the AEM. It also explains why we have spent so much time emphasizing the need to optimize the beam current through use of higher brightness sources and modifying the specimen-detector configuration to maximize the collection angle, while minimizing the various sources of spurious radiation.

> We'll define the minimum detectability in terms of the minimum mass fraction (MMF), which represents the smallest concentration of an element (e.g., in wt.% or ppm) that can be measured in the analysis volume.

Alternatively, the minimum detectable mass (MDM) is sometimes used; the MDM describes the smallest amount of material (e.g., in mg) we can detect. We'll use the MMF, since materials scientists are more used to thinking of composition in terms of wt.% or at.%.

36.7.A. Experimental Factors Affecting the MMF

We can relate the MMF to the practical aspects of microanalysis through the expression of Ziebold (1967)

$$ \text{MMF} \propto \frac{1}{\sqrt{P\frac{P}{B} n\tau}} \qquad [36.14] $$

Here, P is the X-ray count rate in the characteristic peak (above background) of the element of interest, P/B is the peak-to-background count-rate ratio for that peak (defined here in terms of the same width for both P and B), and τ is the analysis time for each of n analyses.

To increase P you can increase the current in the electron beam and/or increase the thickness (t) of the specimen. To increase P/B you can increase the operating voltage (E_0), which is easy, and decrease instrumental contributions to the background, which is not so easy (Zemyan and Williams 1994). Improvements in AEM instrument design, such as using a high-brightness and/or an intermediate voltage source, and a larger collection angle for the XEDS will also increase P. To increase P/B, you need a stable instrument with a clean vacuum environment to minimize or eliminate specimen deterioration and contam-

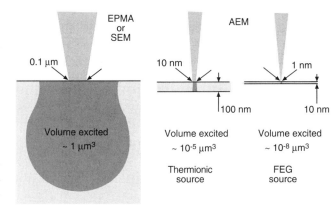

Figure 36.9. Comparison of the relative size of the beam–specimen interaction volumes in an EPMA, a thermionic source AEM, and an FEG-AEM with a bulk, thin, and ultrathin specimen, respectively.

ination. Improved AEM stage design, to minimize stray electrons and bremsstrahlung radiation, both of which contribute background to the detected spectrum, will also help to increase P/B, as we discussed back in Chapter 33.

> Remember, however, that the Fiori definition of P/B is not the one used in Ziebold's equation (36.14) if you actually want to calculate the MMF.

The other variables in equation 36.14 are the time and number of analyses, which are entirely within your control as operator. Usually, both n and τ are a direct function of your patience and a 5–10 min coffee break is usually the maximum time for any one analysis. With computer control of the analysis procedure, however, there should be no limit to the time available for analysis. Particularly when detection of very small amounts of material is sought, τ should be increased to very long times. In the future, a period of several hours or overnight will not be considered unreasonable. Of course, the investment of so much time in a single analysis is dangerous unless you have judiciously selected the analysis region, and you are confident that the time invested will be rewarded with a significant result. Obviously you should minimize factors that degrade the quality of your analysis with time, such as contamination, beam damage, and specimen drift. Therefore, you should only carry out long analyses if your TEM is clean (preferably UHV) and your specimen is also clean and stable under the beam. Any specimen drift must be corrected by computer control during the analysis.

36.7.B. Statistical Criterion for the MMF

We can also define the MMF by a purely statistical criterion. We discussed back in Chapter 34 that we can be sure a peak is present if the peak intensity is greater than three times the standard deviation of the counts in the background under the peak. From this we can come up with a definition of the detectability limit which, when combined with the Cliff–Lorimer equation (assuming Gaussian statistics), gives the MMF (in wt.%) of element B in element A as

$$C_B(\text{MMF}) = \frac{3\left(2\,I_B^b\right)^{\frac{1}{2}}C_A}{k_{AB}\left(I_A - I_A^b\right)} \qquad [36.15]$$

where I_A^b and I_B^b are background intensities for elements A and B, I_A is the raw integrated intensity of peak A (including background), C_A is the concentration of A (in wt.%), and k_{AB}^{-1} is the reciprocal of the Cliff–Lorimer k factor. However, if we express the Cliff–Lorimer equation as

$$\frac{C_A}{k_{AB}\left(I_A - I_A^b\right)} = \frac{C_B}{\left(I_B - I_B^b\right)} \qquad [36.16]$$

and substitute it into equation 36.15, the MMF is

$$C_B(\text{MMF}) = \frac{3\left(2\,I_B^b\right)^{\frac{1}{2}}C_B}{I_B - I_B^b} \qquad [36.17]$$

Experimentally, low count-rates from thin specimens mean that typical values of MMF are in the range 0.1% to 1%, which is rather large compared with some other analytical techniques. The best compromise in terms of improving MMF while maintaining X-ray spatial resolution is to use high operating voltages (300 to 400 kV) and thin specimens to minimize beam broadening. The loss of X-ray intensity, P, a consequence of using thin specimens, can be

compensated in part by the higher voltages and/or by using an FEG where a small spot size of 1 to 2 nm can still be maintained. In summary, optimum MMF and spatial resolution can be obtained by using a high-brightness, intermediate voltage source with thin foils, perhaps of the order of $t \sim 10$ nm. Under these circumstances, MMF values <0.1 wt.% will become routine. Figures 36.9 and 36.10 summarize the classic compromise between resolution and detectability (Lyman 1987).

36.7.C. Comparison with Other Definitions

The MMF definition is not the only way we can measure detectability limits. Currie (1968) has noted at least eight definitions in the analytical chemistry literature. Currie defined three specific limits:

■ The decision limit (L_c): Do the results of your analysis indicate detection or not?
■ The detection limit (L_d): Can you rely on a specific analysis procedure to lead to detection?
■ The determination limit (L_q): Is a specific analysis procedure precise enough to yield a satisfactory quantification?

For I_B counts from element B in a specific peak window and I_B^b in the background, it can be shown that

$$L_C = 2.33\sqrt{I_B^b} \qquad [36.18]$$

$$L_d = 2.71 + 4.65\sqrt{I_B^b} \qquad [36.19]$$

$$L_q = 50\left\{1 + \left(1 + \frac{I_B^b}{12.5}\right)^{\frac{1}{2}}\right\} \qquad [36.20]$$

If there are sufficient counts in the background

$$L_d = 4.65\sqrt{I_B^b} \qquad \text{when } I_B^b > 69 \qquad [36.21]$$

$$L_q = 14.1\sqrt{I_B^b} \qquad \text{when } I_B^b > 2500 \qquad [36.22]$$

Comparison of these definitions with the statistical criterion in the previous section shows that $C_{\text{MMF}} \approx L_d$. So if you want to quantify an element, not just determine that it is present (L_d), then you need substantially more (~3×) of the element in your specimen (Zemyan 1995). Rather than do the experiment yourself, it is possible to simulate spectra from small amounts of element B in A (or vice versa), using DTSA. We recommend that you simulate your analysis before embarking on a time-consuming experiment which may be futile, because the amount of the element you are seeking is below the MMF.

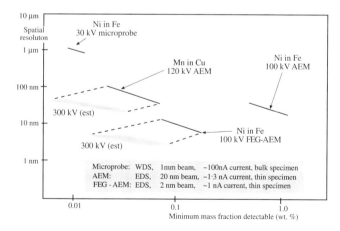

Figure 36.10. Calculation of the relationship between MMF and spatial resolution for the EPMA and a range of AEMs. The inverse relationship between the MMF and resolution is clear, although it is also apparent that the high-brightness sources and high-kV electron beams in the AEM can compensate for the decreased interaction volume in a thin foil.

36.7.D. Minimum Detectable Mass

The MMF values may seem poor compared with other analytical techniques which report ppm or ppb detectability limits. However, it's a different matter if you calculate what the MMF translates to in terms of the minimum detectable mass (MDM).

> The MDM is the minimum number of atoms detectable in the analyzed volume.

Using data for the MMF of Cr in a 304L stainless steel measured in a VG HB-501 AEM with an FEG, Lyman and Michael (1987) obtained an MMF of 0.069 wt.% Cr in a 164-nm foil with a spatial resolution of 44 nm and a 200-s counting time. The electron beam size was 2 nm (FWTM) with a beam current of 1.7 nA. In this analysis, an estimated 2×10^4 atoms were detected and the MDM was less than 10^{-19} g. If the counting time is increased by a factor of 10 and the operating voltage is increased to 300 kV, the spatial resolution would improve to ~15 nm and the MMF would improve to ~0.01 wt.%. Thus about 300 atoms could be detected. For a foil thickness of 16 nm (1/10th the above measured thickness), the MMF would degrade to ~0.03 wt.%. However, the spatial resolution would improve to about 2 nm. For this case, about 20 atoms would be detected corresponding to less than 10^{-22} g, which is an amazing figure by any standards. Therefore in ~10-nm-thick specimens, with a spatial resolution approaching the beam diameter d, of 1 to 2 nm, we will be able to detect the presence of 10 to 100 atoms in the analysis volume (10^{-8} μm^3); preliminary data reporting <10 atoms have been published (Lyman et al. 1994). For comparison, in the EPMA with 1 μm^3 excitation volume and a 0.01 wt.% MMF, ~10^7 atoms are detected in the analysis volume.

CHAPTER SUMMARY

You cannot optimize spatial resolution and minimum detectability in the same experiment. You must decide which of the two criteria is more important for the result you're seeking:

■ To get the best spatial resolution, operate with the thinnest foils and the highest-energy electron beam. Use an FEG if possible.

■ To measure the specimen thickness, use the parameterized EELS approach, otherwise choose between any of the several X-ray intensity methods, or CBED for a crystalline foil.

■ To get the best MMF, use the brightest electron source, or the largest beam and thickest specimen, and count for as long as possible.

■ If you want the best resolution *and* MMF, an FEG is essential, along with a clean specimen and computer-controlled drift correction; patience is also desirable.

REFERENCES

General References

Berriman, J., Bryan, R.K., Freeman, R., and Leonard, K.R. (1984) *Ultramicroscopy* **13**, 351.
Goldstein, J. I., Williams, D.B., and Cliff, G. (1986) in *Principles of Analytical Electron Microscopy* (Eds. D.C. Joy, A.D. Romig Jr., and J.I. Goldstein), p. 155, Plenum Press, New York.
Scott, V.D. and Love, G. (1987) *Mat. Sci. Tech.* **3**, 600.

Specific References

Currie, L.A. (1968) *Anal. Chem.* **40**, 586.
Goldstein J.I., Costley, J.L., Lorimer, G.W., and Reed, S.J.B. (1977) *Scanning Electron Microscopy*, **1** (Ed. O. Johari), p. 315, IITRI, Chicago, Illinois.

Heinrich, K.F.J., Newbury, D.E., and Yakowitz, H., Eds. (1975) NBS Special Publication 460, U.S. Dept. of Commerce, Washington, D.C.
Horita, Z., Ichitani, K., Sano, T., and Nemoto, M. (1989) *Phil. Mag.* **A59**, 939.
Joy, D.C. (1995) *Monte Carlo Modeling for Electron Microscopy and Microanalysis,* Oxford University Press, New York.
Lorimer, G.W., Cliff, G., and Clark, J.N. (1976) in *Developments in Electron Microscopy and Analysis* (Ed. J.A. Venables), p. 153, Academic Press, London.
Lyman, C.E. (1987) in *Physical Aspects of Microscopic Characterization of Materials* (Eds. J. Kirschner, K. Murata, and J.A. Venables), p. 123, *Scanning Microscopy International,* AMF O'Hare, Illinois.
Lyman, C.E. and Michael, J.R. (1987) in *Analytical Electron Microscopy-1987* (Ed. D.C. Joy), p. 231, San Francisco Press, San Francisco, California.

Lyman, C.E., Goldstein, J.I., Williams, D.B., Ackland, D.W., Von Harrach, S., Nicholls, A.W., and Statham, P.J. (1994) J. *Microsc.* **176**, 85.

Michael, J.R., Williams, D.B., Klein, C.F., and Ayer, R. (1990) *J. Microsc.* **160**, 41.

Porter, D.A. and Westengen, H. (1981) in *Quantitative Microanalysis with High Spatial Resolution* (Eds. M.H. Jacobs, G.W. Lorimer, and P. Doig), p. 94, The Metals Society, London.

Reed, S.J.B. (1982) *Ultramicroscopy* 7, 405.

Williams, D.B., Michael, J.R., Goldstein, J.I., and Romig, A.D. Jr. (1992) *Ultramicroscopy* **47**, 121.

Zemyan, S.M. (1995) Ph.D. dissertation, Lehigh University.

Zemyan, S.M. and Williams, D.B. (1994) *J. Microsc.* **174**, 1.

Ziebold, T.O. (1967) *Anal. Chem.* **39**, 858.

Electron Energy-Loss Spectrometers

37

CHAPTER PREVIEW

Electron energy-loss spectrometry (EELS) is the analysis of the energy distribution of electrons that have interacted inelastically with the specimen. These inelastic collisions tell us a tremendous amount about the electronic structure of the specimen atoms, which in turn reveals details of the nature of these atoms, their bonding and nearest-neighbor distributions, and their dielectric response. In order to examine the spectrum of electron energies we almost invariably use a magnetic prism spectrometer which, when interfaced to a TEM, creates another form of AEM. The magnetic prism is a simple, but highly sensitive, device with resolving power of approximately 1 eV even when the energy of the incident electron beam is up to 400 keV. Despite its simplicity the magnetic prism is operator-intensive and there is not yet the degree of software control to which we are accustomed with XEDS. In this chapter we'll describe the operational principles, how to focus and calibrate the spectrometer, and how to determine the collection semiangle (β). This angle is a most important parameter for interpreting your experimental data. In subsequent chapters we'll go on to look at the spectra, the information they contain, and how we extract quantitative data and images from them. As with XEDS there are standard tests to determine that the spectrometer is working correctly, and we'll describe these also.

As a word of encouragement, or warning, you may get the impression from reviewing the older literature that EELS is the study of small blips which can only be seen by the "trained" eye. While the blips are still often small, we can now be very confident of our interpretation of these spectra. The EELS technique has come to be an excellent complement to the more widely used X-ray spectrometry, since it is well suited to the detection of light elements which are difficult to analyze with XEDS.

Electron Energy-Loss Spectrometers

37

37.1. WHY DO ELECTRON ENERGY-LOSS SPECTROMETRY?

When the electron beam traverses a thin specimen, it loses energy by a variety of processes that we first discussed way back in Chapter 4. The reason we do EELS is so we can separate these inelastically scattered electrons and quantify the information they contain. We've already seen contrasting aspects of inelastic scattering in the TEM:

- Kikuchi lines occur in DPs; these electrons in Kikuchi lines are diffracted at precisely the Bragg angle, and give us much more accurate crystallographic information than the SAD pattern. In thick specimens, many of the electrons in Kikuchi lines are inelastically scattered.
- Chromatic aberration due to energy-loss electrons following different paths through the lenses limits the TEM image resolution, although you can avoid this by using very thin specimens, or STEM imaging. We'll see also, in Chapter 40, that EELS can filter out the chromatic aberration effect in TEM images.
- Specimen damage, which is of course undesirable, is often caused by inelastic interactions.

After reading the next three chapters you should agree that EELS is useful too. If you have an energy-loss spectrometer, it may be that inelastic scatter, in general, is something you would like to happen in your specimens.

The technique of EELS predates X-ray spectrometry; if you want to read a brief history of the technique see the book by Egerton (1996), which we will refer to on many occasions. In fact, the experimental pioneers of EELS, Hillier and Baker (1944), were the same two scientists who first proposed and patented the idea of X-ray spectrometry in an electron-beam instrument similar to the EPMA. In contrast to X-ray analysis, EELS has been very slow to develop and still remains firmly in research laboratories rather than applications laboratories. Since probe-forming AEMs became widespread, EELS has become more popular, primarily because it complements XEDS through better detection of the light elements. But as you'll see, we can extract a lot more from the spectra than merely elemental identification. When you have finished this set of chapters you will be ready to read Egerton's text and you will find the book edited by Disko *et al.* (1992) to be another excellent review. Two special issues of *Microscopy, Microanalysis, Microstructure* (Krivanek 1991, 1995a) and *Ultramicroscopy* (Krivanek 1995b) contain papers from EELS workshops.

Currently, there are only two major manufacturers of electron spectrometers for TEMs and they produce radically different instruments, designed for different purposes. We'll describe these two types of electron spectrometers in some detail.

> Two types of commercial spectrometers are presently manufactured: the magnetic prism spectrometer (Gatan) and the omega filter [Zeiss (now LEO)].

The magnetic prism is designed with energy spectrometry as its primary function; this application will constitute the bulk of this chapter. The omega filter is used mainly for energy-filtered imaging although spectra can be obtained; it is a specialized technique because this spectrometer has to be built into the microscope column rather than an optional addition (see later in Section 37.6). However, the Gatan Image Filter (GIF) may soon change this situation, particularly in the materials sciences, since the GIF combines both spectral and imaging capabilities. Magnetic electron spectrometers, along with electrostatic or combined electrostatic and magnetic systems, have been the subject of reviews by Metherell (1971) and Egerton (1996). If you're an instrument enthusiast you should read these articles. Perhaps because there are so few manufacturers of the spectrometers, the competitively driven progress in user-friendly instrumentation and software control that has pushed X-ray spectrometry forward over the last two decades has been slow to occur in EELS; this lack of user friendliness, in part, accounts for the relatively small number of users of the technique.

> You should know also that there is another area of EELS research which uses electron spectrometers to measure exceedingly small (millivolt) energy losses in low-energy electron beams reflected from the surfaces of samples in UHV surface-chemistry instrumentation such as ESCA and Auger systems. We will ignore this type of EELS completely and deal only with transmission EELS studies of high-voltage electron beams.

There are two fundamentally different ways of detecting the spectrum generated by the magnetic prism spectrometer: either serially or, more efficiently, in parallel (see Section 37.3). Also, you can operate your TEM either in image mode or diffraction mode (see Section 37.4) and this choice has a major effect on the information that can be gathered. Before we look at these options, however, we need to look at the magnetic prism spectrometer itself.

37.2. THE MAGNETIC PRISM; A SPECTROMETER AND A LENS

We use a magnetic prism rather than one of the other kinds of spectrometer (e.g., electrostatic) for several reasons:

- It is compact, and therefore easily interfaced to the TEM. (Remember the WDS problem.)
- It offers sufficient energy resolution to distinguish all the elements in the periodic table and so is ideal for microanalysis.
- Electrons in the energy range 100–400 keV, typical of AEMs, can be dispersed sufficiently to detect the spectrum electronically, without limiting the energy resolution.

Schematic diagrams of the spectrometer optics are shown in Figures 37.1A and B. A picture of a Gatan spectrometer, which has to be installed beneath the camera system of a TEM or after the ADF detector in a DSTEM, is shown in Figure 37.2. Because these spectrometers are so widespread, many of the numerical values in this chapter are

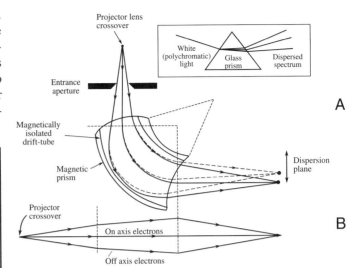

Figure 37.1. Ray paths through a magnetic prism spectrometer showing (A) dispersion and focusing of the electrons in the plane of the spectrometer and (B) the lens-focusing action in the plane normal to the spectrometer; compare the nonfocusing action of a glass prism on visible light (inset).

taken from the Gatan literature. For the details of operation you should, of course, read the instruction manual.

From Figure 37.1A we can see that electrons are selected by a variable entrance aperture (diameters: 1, 2, 3, or 5 mm in the Gatan system). The electrons travel down a "drift tube" through the spectrometer and are deflected through ≥ 90° by the surrounding magnetic field. Electrons with greater energy loss (dashed line) are deflected further than those suffering zero loss (full line). A spectrum is thus formed in the dispersion plane which consists of a distribution of electron counts (again incorrectly referred to as intensity) (I) versus energy loss ($\mathcal{E}$). This process is exactly analogous to the dispersion of white light by a glass prism.

> Although we've consistently used the letter E for energy, energy loss should, therefore, be denoted by ΔE, since it's a change in energy. However, it is a convention in the EELS literature to use E interchangeably for both an energy loss (e.g., the plasmon loss E_p) and a specific energy (e.g., the critical ionization energy E_c). As a compromise we will use $\mathcal{E}$ (note the different font), but remember it really means a change in E.

Now if you look at Figure 37.1B, you'll see that electrons suffering the same energy loss but traveling in both on-axis and off-axis directions are also brought back to a focus in the dispersion plane of the spectrometer,

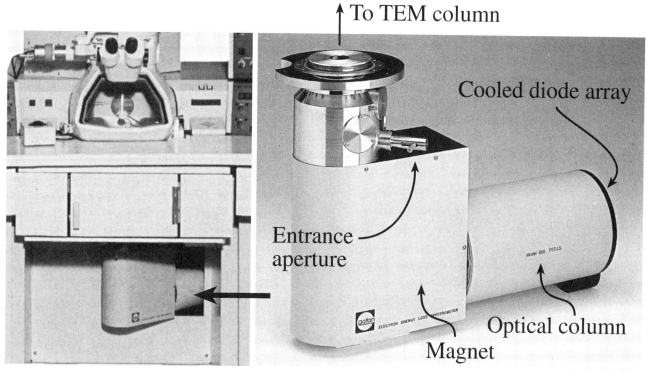

Figure 37.2. A Gatan parallel-collection magnetic prism spectrometer interfaced below the viewing screen of an AEM showing the shielded container for the magnet and the optical column and diode-array detection system.

which thus acts as a magnetic lens. The object plane of the spectrometer is usually set at the projector lens back focal plane, coincident with the differential pumping aperture. This focusing action is not seen in the otherwise analogous glass prism. Many examples of spectra will be given in the next two chapters.

37.2.A. Focusing the Spectrometer

Because the spectrometer is also a lens, you have to know how to focus it, and how to minimize the aberrations and astigmatism that are inherent in any magnetic lens. Correction of second-order aberrations and astigmatism are minor steps which we will not describe, but focusing is rather important.

The spectrometer has to focus the electrons because off-axis electrons experience a different magnetic field than on-axis electrons. The spectrometer is an axially *asymmetric* lens unlike the other TEM lenses. The path length of off-axis electrons through the magnet also varies, and the magnet has to be carefully constructed to ensure correct compensation for different electron paths so that focusing occurs. This correction is achieved by machining the entrance and exit faces of the spectrometer so they are not normal to the axial rays,

as shown in Figure 37.1A. These nonnormal faces also act to ensure that electrons traveling out of the plane of the paper in Figure 37.1A are also focused in the dispersion plane, as shown in Figure 37.1B. Such a spectrometer is described as "double focusing." The faces are also curved to minimize aberrations.

As with any lens, the spectrometer takes electrons emanating from a point in an object plane and brings them back to a point in the image (dispersion) plane. Because the spectrometer is an asymmetric lens, we have to fix both the object distance and image distance if we want to keep the spectrum in focus. The object plane of the spectrometer depends on the TEM you are using.

> In a TEM/STEM, or a DSTEM with post-specimen lenses, the object plane is the back focal plane of the projector lens.
> In DSTEMs with no post-specimen lenses, the object plane is the plane of the specimen.

In the TEM the projector lens setting is usually fixed, and the manufacturer has already set this plane to concide with the differential pumping aperture separating the column from the viewing chamber. In some dedicated DSTEMs there are no post-specimen lenses, so the object plane of the spectrometer must be the plane of the specimen. In this case it is essential that you keep the specimen height constant.

Now in practice, the back focal plane of the projector does move a little as you change operating modes (for example, from TEM to STEM) and so you have to be able to adjust the spectrometer. You do this by looking at the electrons that come through the specimen without losing any energy. These electrons have a Gaussian-shaped intensity distribution which we call the "zero-loss peak"; we'll talk about this more in the next chapter. You can see the zero-loss peak on the CRT or computer display of the EELS system. You focus the peak by adjusting a pair of pre-spectrometer quadrupoles until it has a minimum width and maximum height. To correct second-order effects, a pair of sextupoles is also available. The actual method of focusing depends on the type of spectrometer. In a parallel-collection system, focusing is controlled directly by the quadrupoles. In a serial-collection system (see below), there is a slit in the dispersion plane. There are some rules to guide you when adjusting the slits:

- If the slit is too wide, then the zero-loss peak has a flat top and is very broad.
- If the slit is too narrow, you lose spectral intensity.
- If your spectrometer is not focused on the slit, the peak is also broad, but more Gaussian-shaped, as in Figure 37.3.

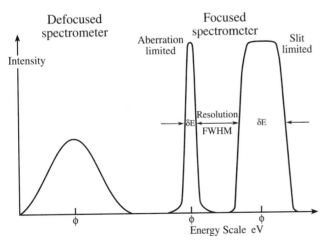

Figure 37.3. The zero-loss peak in a spectrum showing the effect of a defocused spectrometer, and the slit-limited condition, i.e., slits too wide (for serial collection only). In the usual focused condition, the resolution of the spectrometer is defined as the FWHM of the peak.

37.2.B. Calibrating the Spectrometer

We calibrate the spectrometer by placing an accurately known voltage on the drift tube. You will see this voltage displace the spectrum by a fixed amount. Alternatively, as in XEDS, you can look for features in a spectrum from a known specimen that occur at specific energies spanning the spectral display range, such as the zero-loss peak (at 0 eV) and the Ni L edge at 855 eV. (See Figure 37.4 and the next chapter for more details on the spectrum itself.) Modern electronics are reasonably stable and the calibration doesn't shift substantially, but you should check it regularly throughout an operating session since shifts of a few eV do occur and these are of the same order as the energy resolution of the spectrometer.

37.3. ACQUIRING A SPECTRUM

Figure 37.4A shows an EELS spectrum, which we'll describe in detail in the next chapter. For the time being note:

- The zero-loss peak is very intense.
- The intensity range is enormous; this graph uses a logarithmic scale.

There are two ways in which we acquire the spectrum. We either build up the spectrum one channel at a time, which is known as serial-acquisition EELS, or SEELS, or we acquire all the channels simultaneously, which is parallel-acquisition EELS, or PEELS (the word "acquisition" is often omitted when discussing the subject). The two modes are shown in the schematic spectra in Figure 37.4B and C. The intensity changes shown correspond to ionization "edges" and each element exhibits characteristic edges at specific values of $\mathcal{E}$. For example, the carbon K edge onset at 284 eV corresponds to the critical energy E_C required to eject the carbon K shell electron; more about this in Chapter 38.

37.3.A. Serial Collection

SEELS is the original, but least efficient, method of acquisition. No serial systems are now manufactured, but many remain in TEM labs. As shown in Figure 37.5A, the spectrometer system scans or "ramps" the spectrum across a slit in the dispersion plane, leaving the spectrum for a fixed "dwell time" (τ) at each energy loss. The collection and measurement is quite straightforward.

- Ramping is achieved magnetically by changing the current through a set of coils placed after the spectrometer magnet.

A

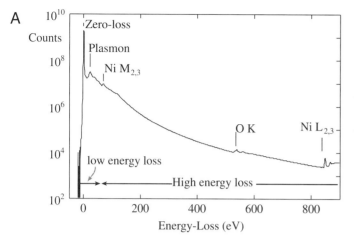

B

C

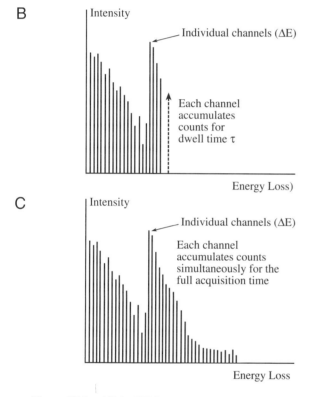

Figure 37.4. (A) An EELS spectrum displayed in logarithmic intensity mode. The zero loss is an order of magnitude more intense than the low-energy-loss portion, which is itself many orders of magnitude more intense than the small ionization edges, identified in the high-energy-loss range. The spectra can be acquired (B) serially or (C) in parallel. In serial collection, each of the energy channels accumulates counts for a given dwell time τ (typically 100 ms) before the next channel (energy range) is selected. In parallel collection, the complete spectrum is gathered simultaneously in all the available channels.

■ A plastic scintillator located directly behind the slit receives an electron flux in time τ.

■ The integrated electron current is converted to photons and amplified by a photomultiplier (PM) tube.

■ After each dwell time, the total signal from the PM is then assigned to a channel corresponding to a specific energy loss, $\mathcal{E}$, in a multichannel analyzer (MCA).

We've already discussed scintillator-PM systems of the sort used for SE detection in the SEM or STEM, or direct beam (BF) detectors in the STEM (see Chapter 7).

■ They have a tremendous gain and therefore can handle a large intensity range.

■ They show a rapid response to intensity changes and exhibit both low noise and a high detective quantum efficiency, or DQE.

The DQE is close to 1 under ideal conditions, although the absolute DQE of SEELS is very low (≈0.001) since most of the electrons are wasted at any single acquisition time. These factors are particularly important in EELS, because the spectrum intensity can vary by many orders of magnitude. However, the susceptibility of scintillators to beam damage also means that the intense low-energy part of the EELS spectrum can be a health danger to the scintillator. The Gatan systems give an unmistakable audible warning when the electron flux is too great. Even with careful operation the plastic scintillator in your SEELS will become damaged and, as a wise precaution, you should replace it every few months.

The spectrum display is built up in a serial manner, as we showed in Figure 37.4B. Typically, the MCA will have 1024 or 2048 channels. The display resolution can be selected from < 0.1 eV/channel to about 10 eV/channel, depending on how much of the spectrum you want to gather. For example, if you set the display resolution to 1 eV per channel, the entire EELS signal out to 1024 eV energy loss will be recorded. But even if you choose a short τ, e.g., 100 ms, it will still take you about 100 s to accumulate a full spectrum out to 1024 eV, and in 100 ms the total intensity in each channel is going to be small. In particular, the limited energy range that may be of interest to you will only have been sampled for a few seconds at best. So when you know what portion of the spectrum is of interest, you should restrict collection to fewer channels for a longer τ, or accumulate several spectra and add the intensities together in order to get satisfactory counting statistics across the full spectrum. We'll see later that you can influence the total intensity by your choice of operating mode.

During acquisition of a spectrum, the intensity changes by several orders of magnitude (see Figure 37.4A), and usually we are interested primarily in the low-intensity (high-energy-loss) part of the spectrum. Therefore, we have

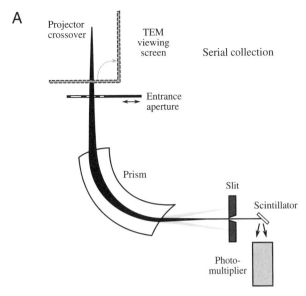

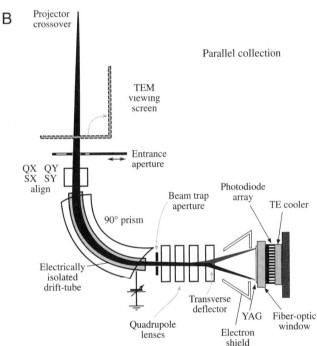

Figure 37.5. Comparison of EELS acquisition modes. (A) Serial collection of the spectrum through a slit onto a scintillator-PM system. (B) Parallel collection onto a YAG scintillator fiber-optically coupled to a thermoelectrically cooled (TE) semiconductor diode array. (Q = quadrupole, S = sextupole)

to change the display scale during SEELS spectrum collection in order to present a visible display of the low-intensity part of the spectrum. We can do this in one of two ways: either we increase τ by a factor of 10–100, or we electronically increase the gain of the scintillator-PM system by

some orders of magnitude and operate it in two different modes, as explained below.

For very high electron fluxes encountered in the zero-loss and low-energy-loss portion of the spectrum, we use "analog collection." The total voltage generated at the exit of the PM can be related directly to the total electron current incident on the scintillator. Using a voltage-to-frequency (V/F) converter, this voltage is converted to a pulse that can be fed directly to the computer display. By changing the gain of the V/F converter we can handle electron fluxes in excess of 10^{10}/s (about 1 nA) with no problem. At this kind of incident current, the individual pulses overlap and are all integrated to give a continuous output. At the high-energy-loss end of the spectrum, where electron intensity is very low, the collection system can be changed to count the photon bursts generated by each electron, and this is termed "single-electron counting." You throw a switch on the SEELS electronics control panel to change the detection mode at a given point (energy loss) in the serial collection, or you can change modes under software control, usually at the same energy in the spectrum where the display gain change occurs.

We will see that it is sometimes essential to record the spectrum over a wide range, say from zero to several hundred eV loss. In this case, the intense low-loss region must be gathered and displayed in the same spectrum as the less-intense high-loss region. A potential problem in this situation is that, even if the intense low-loss region does not physically damage the detector, it may cause it to glow and the glow persists for some time after the electrons hit the scintillator. So it is possible that while you are gathering the low-intensity part of the spectrum, the afterglow will contribute unwanted noise.

> To avoid the afterglow, you should acquire all SEELS spectra in reverse-scan fashion. Start at the high loss, then go to the low-loss region, and finish at zero.

If the zero-loss peak is not required, then it is good practice to cease acquisition at about 5–10 eV to save the scintillator from the accumulated effects of high electron fluxes.

37.3.B. Parallel Collection

PEELS gathers the whole energy spectrum simultaneously and is much more efficient than SEELS. PEELS comprises a YAG scintillator coupled via fiber optics to a semiconductor photodiode array in the dispersion plane of the spectrometer, as shown in Figure 37.5B. The array consists of

1024 electrically isolated and thermoelectrically cooled silicon diodes, each about 25 mm across. These diode arrays show varying responses and exhibit specific artifacts, which we'll discuss in the next chapter.

The resultant spectrum accumulates across the whole energy range simultaneously, as we showed schematically back in Figure 37.4C. Rather than having a dwell time as in SEELS, we now have an integration time which can vary from a few msec to several hundred seconds. After integration, the whole spectrum is read out via an amplifier through an A/D converter and into an MCA system. Reasonable spectra can be acquired in a fraction of a second, making PEELS imaging a practical reality. We'll see more about this in Section 40.3.

> The advantage of PEELS is that all regions of interest are gathered for the whole integration time, and not just some fraction of the acquisition time as in SEELS. Thus PEELS is much more efficient than SEELS, and its DQE is ~0.5.

A quick warning: you can damage the YAG scintillator, particularly in intermediate-voltage microscopes. Ways to avoid this problem, especially for the intense zero-loss beam, are still being developed. Currently, the zero-loss beam intensity is attenuated or deflected off the scintillator if it is not required, or if the beam current exceeds 0.5 nA. While the SEELS system can handle intense signals, the PEELS diode array saturates at signal intensities of about 16,000 counts.

> You must select an integration time that won't saturate the detector and then collect as many consecutive integrations as you need.

One other advantage of the PEELS system is that the scintillator shield is designed to act as a Faraday cup, and so you can use it to measure the total beam current. Usually, the beam is moved onto the shield whenever acquisition ceases, so a constant record of the beam current is available.

To summarize:

- ■ SEELS detects one channel at a time; the detector is easy to optimize and simple to operate.
- ■ PEELS detects the whole spectrum at one time, but the diode array is hard to optimize.
- ■ PEELS exhibits artifacts, and has more complex electron optics, but is 2–3 orders of magnitude more efficient than SEELS with a relatively high DQE.

37.3.C. Spectrometer Dispersion

We define the dispersion as the distance in the spectrum (dx) between the positions of electrons differing by energy dE. It is a function of the strength of the magnetic field (which is governed by the strength (i.e., size) of the spectrometer magnet) and the energy of the incident beam, E_0. In the commercial serial spectrometers, the radius of curvature (R) of electrons traveling on axis is about 200 mm, and for 100-keV electrons dx/dE is about 2 µm/eV. This dispersion, while small, is sufficiently large so as not to limit the energy resolution (see below). A serial-detection system can process the spectrum without any post-spectrometer magnifying lenses. For parallel collection this dispersion value is inadequate, and typically electrons with an energy range of about 15 eV would fall on each 25-µm-wide diode. Therefore, the dispersion plane has to be magnified ~15× before the spectrum can be detected with resolution closer to 1eV. This magnification requires post-spectrometer lenses; 4 quadrupoles are used in the Gatan system. The dispersion should be linear across the diode array; you can check this by measuring the separation of a known pair of spectral features (e.g., zero loss and C K edge) as you displace the spectrum across the diode array.

You may wonder why we don't record the spectrum on film rather than collect it electronically. In fact, this was the first method used to detect electron spectra, but it is an analog method and photographic film does not have a linear response over the usual range of spectral intensities. Furthermore, the grain size of the photographic emulsion (10–20 µm) would limit energy resolution to about 5 eV unless the dispersion were increased, so photographic recording is no longer used.

37.3.D. Spectrometer Resolution

We define the energy resolution of the spectrometer as the FWHM of the zero-loss peak (see back in Figure 37.3). If you don't focus your spectrometer as we just described, then you won't get the best resolution. The best resolution you can get is determined by the type of electron source. As we discussed back in Chapter 5 (see Table 5.1), at ~100 keV a W source has the worst energy resolution (2.5 eV), and a LaB$_6$ is slightly better than W at 1.5 eV while a cold FEG gives the best value (0.3 eV). Because of the high emission current from thermionic sources, the energy resolution is in fact limited by electrostatic interactions between electrons at the filament crossover. This electron–electron interaction is called the Boersch effect. You can partially overcome this by undersaturating the filament and using only the electrons in the halo. Under these circumstances a LaB$_6$ source can attain a resolution below

1 eV, but at the expense of a considerable loss of signal, for which you can compensate by increasing the beam size and/or the C2 aperture.

> The energy resolution decreases slightly as the energy loss increases, but it should be no worse than ~1.5 times the zero-loss peak width up to 1000-eV energy loss.
> If you operate at higher voltage, you should also expect a degradation of energy resolution as the kV increases, approximately tripling from 100 kV to 400 kV.

Because the magnetic prism is so extremely sensitive, external magnetic fields in the microscope room may limit the resolution. You may see a disturbance to the spectrum directly if you sit in a metal chair and move around, or if you open metal doors into the TEM room. Remember, for comparison, that the energy resolution of XEDS is >100 eV.

For different EELS operations, different factors affect the resolution. In a SEELS system the resolution is most easily changed by adjusting the slit width. A larger slit width results in poorer energy resolution, but has the advantage of increasing the total current into the detection system. In a PEELS system optimum resolution requires a small projector crossover and a small (1 mm or 2 mm) entrance aperture. The resolution is degraded by choosing a larger entrance aperture because of off-axis beams; i.e., degradation is caused by a C_s effect for the lenses in the energy analyzing spectrometer. Similarly, the resolution may change as you deflect the zero-loss peak onto different regions of the photodiode, although this should not happen if the spectrometer optics are properly aligned.

37.3.E. Point-Spread Function

In a PEELS, you can reduce the magnification of your spectrum so that the zero-loss peak occupies only a single photodiode channel. Any intensity registered outside that single channel is an artifact of the detector system array and is called the point-spread function. This function acts to degrade the inherent resolution of the magnetic spectrometer. The zero-loss peak may spread on its way through the YAG scintillator and the fiber optics before hitting the photodiode. Figure 37.6 shows the point-spread function of a PEELS and clearly there is intensity well outside a single channel. This is important because this spreading broadens features in your spectrum, such as fine structure in ionization edges, and you need to remove it by deconvolution (see Section 39.6). The concept is essentially the same as the point-spread function we discussed for HRTEM.

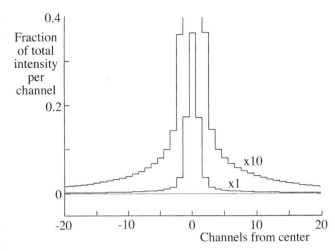

Figure 37.6. The point-spread function, showing the degradation of the intense well-defined zero-loss peak through spreading of the signal as it is transferred from the scintillator via the fiber-optic coupling to the photodiodes. The peak should occupy a single channel but is spread across several channels.

37.4. IMAGE AND DIFFRACTION MODES

When performing EELS in a TEM/STEM, you can operate in either of two modes, and the terminology for this is confusing. If you operate the TEM such that an image is present on the viewing screen, then the back focal plane of the projector lens contains a DP, and the spectrometer uses this pattern as its object. From the spectroscopists' viewpoint, therefore, this is termed "diffraction mode" or "diffraction coupling," but from the microscopists' viewpoint it is more natural to call this "image mode" since you are looking at an image on the screen. Conversely, if you adjust the microscope so a DP is projected onto the screen (which includes STEM mode in a TEM/STEM), then the spectrometer object plane contains an image, and the terminology is reversed.

> The spectroscopist uses the term "image mode" or "image coupling" and the microscopist says "diffraction mode."

Both sets of terms appear in the literature, often without precise definition, so it can be rather confusing.

> In this text "image mode" means an image is present on the TEM screen; i.e., we use the microscopists' terminology.

So your first step is to ensure that a focused image or DP is present on your TEM screen, and then the spectrum can be focused onto the dispersion plane.

37.4.A. Spectrometer Collection Angle

The collection semiangle of the spectrometer (β) is the most important variable in quantification, so you should know β for all your standard operating situations. If you do gather spectra with different β, it is difficult to make sensible comparisons without considerable post-acquisition processing. The detailed intensity variations in the spectrum depend on the range of electron scattering angles which are gathered by the spectrometer. Under certain circumstances, the effective value of β can be modified by the beam-convergence semiangle, α, but we'll discuss that when we talk about quantification in Chapter 39.

> β is the semiangle subtended at the specimen by the entrance aperture to the spectrometer.

This definition is illustrated in Figure 37.7. The value of β is affected by the mode of microscope operation, and so we will describe how to measure β under different conditions that may be encountered.

Dedicated STEMs. In a basic DSTEM the situation is straightforward if there are no post-specimen lenses because, as shown in Figure 37.7, the collection angle can be calculated from simple geometry. Depending on the diameter (*d*) of the spectrometer entrance aperture and the distance from the specimen to the aperture (*h*), β (in radians) is given by

$$\beta \approx \frac{d}{2h} \qquad [37.1]$$

This value is approximate and assumes β is small. Since *h* is not a variable, the range of β is controlled by the number and size of available apertures. Therefore, if *h* is ~100 mm, then for a 1-mm-diameter aperture, β is 5 mrads. If there are post-specimen lenses and apertures, the situation is similar to that in a TEM/STEM, as discussed below.

TEM-image mode. Remember that in image mode, a magnified image of the specimen is present on the viewing screen and the spectrometer object plane contains a DP. In contrast to what we just described for a dedicated STEM, the angular distribution of electrons entering the spectrometer aperture below the center of the TEM screen is *independent* of the entrance aperture size. This is because you can control the angular distribution of electrons contributing to any TEM image by the size of the objective aperture in the DP in the back focal plane of the objective lens. If you don't use an objective aperture, then the collection semiangle is very

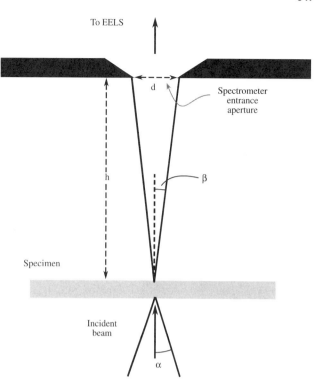

Figure 37.7. Schematic diagram showing the definition of β in a DSTEM in which no lenses exist between the specimen and the spectrometer entrance aperture.

large (>~100 mrads) and need not be calculated accurately, because we'll see that small differences in a large β value do not affect the spectrum or subsequent quantification.

If, for some reason, you do wish to calculate β in image mode with no aperture inserted, you need to know the magnification of the DP in the back focal plane of the projector lens (which is the front focal plane of the spectrometer). This magnification may be described in terms of the camera length *L* of the DP, and this is given by

$$L \approx \frac{D}{M} \qquad [37.2]$$

where *D* is the distance from the projector crossover to the recording plane and *M* is the magnification of the image in that plane. So if *D* is about 500 mm and the screen magnification is 10,000×, then *L* is 0.05 mm. Thus we can show that

$$\beta \approx \frac{r_0}{L} \qquad [37.3]$$

where r_0 is the maximum radius of the DP in the focal plane of the spectrometer. Typically, r_0 is approximately 5 μm, and so β is 0.1 rads or 100 mrad which, as we just said, is so large that we rarely need to know it accurately. In fact, in TEM-image mode without an objective aperture, if you just assume $\beta = 100$ mrad any calculation or quantification you do will be independent of β.

If you insert an objective aperture and you know its size and the focal length of the objective lens, then β can easily be calculated geometrically. To a first approximation, in a similar manner to equation 37.1 above, β is the objective aperture diameter divided by twice the focal length of the objective lens, as shown in Figure 37.8. For example, with a focal length of 3 mm and a 30-μm aperture, β is about 5 mrad.

If you insert an objective aperture, a normal BF image can be seen on the TEM screen and the information in the spectrum is related (with some considerable error) to the area of the image that sits directly above the spectrometer entrance aperture. We will return to this point in more detail in Section 37.4 when we discuss the spatial resolution of microanalysis. Remember also that with the objective aperture in, you cannot do XEDS. Therefore, simultaneous EELS and XEDS is not possible in this mode.

TEM/STEM diffraction mode. In diffraction (also STEM) mode, the situation is a little more complicated. Remember, the object plane of the spectrometer (the projector lens BFP) contains a low-magnification image of the specimen; so you see a DP on the screen and the same DP is in the plane of the spectrometer entrance aperture. Under these circumstances we control β by our choice of the spectrometer entrance aperture, as shown in Figure 37.9.

> If a small objective aperture is inserted, it is possible that it may limit β; the effective value of β at the back focal plane of the projector lens is β/M, where *M* is the magnification of the image in the back focal plane of the projector lens.

You have to calibrate β from the DP of a known crystalline specimen, as also shown in Figure 37.9. Knowing the size of the spectrometer entrance aperture, the value of β can be calibrated by twice the Bragg angle, $2\theta_B$, that separates the 000 spot and a known $hk\ell$, disk. If the effective aperture diameter in the recording plane is d_{eff} and the distance b is related to the angle $2\theta_B$, as shown in Figure 5.8, so

$$\beta \approx \frac{d_{eff}}{2}\frac{2\theta_B}{b}   \qquad [37.4]$$

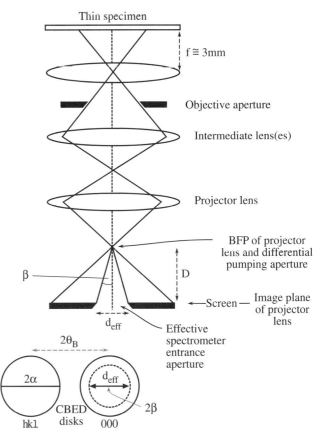

Figure 37.9. The value of β in TEM/STEM diffraction mode is determined by the effective diameter of the spectrometer entrance aperture (d_{eff}), projected into the plane of the diffraction pattern. The value of the d_{eff} can be calibrated by reference to a known diffraction pattern (below) in which d_{eff} can be related to $2\theta_B$.

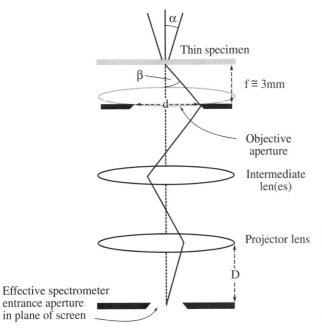

Figure 37.8. The value of β in TEM-image mode is governed by the dimensions of the objective aperture.

The effective entrance aperture diameter d_{eff} at the recording plane is related to the actual diameter d by

$$d_{eff} = \frac{d\,D}{D_A}$$ [37.5]

where D is the distance from the projector crossover to the recording plane (remember, the film is not at the same height as the screen); D_A is the distance between the crossover and the actual entrance aperture. Alternatively, β can be determined directly if the camera length on the recording plane (L) is known, since

$$\beta = \frac{D}{D_A}\frac{d}{L}$$ [37.6]

D_A is typically 610 mm for most Gatan PEELS systems, but D varies from microscope to microscope; you have control over d and L. For example, if D is 500 mm and L is 800 mm, then for the 5-mm entrance aperture, β is ~5 mrads.

If you choose a camera length such that the image of the specimen in the back focal plane of the spectrometer is at a magnification of 1×, then, in effect, you have moved the specimen to the object plane of the spectrometer. This special value of L is equal to the D, which you should know for your own microscope. Then β is simply the entrance aperture diameter divided by D_A (610 mm). Life will be much easier when all these calculations are incorporated in ELP (see Section 1.5).

In summary, the collection angle is a crucial factor in EELS. Large collection angles will give high intensity in the spectrum. If you collect your spectrum in image mode without an objective aperture, then you don't compromise your energy resolution. If you're in diffraction mode and you control β with the entrance aperture, then a large aperture (high intensity, large β) will degrade the energy resolution.

37.4.B. Spatial Selection

Depending on whether you're operating in image or diffraction mode, you obtain your spectrum from different regions of the specimen. In TEM-image mode, you position the area to be analyzed on the optic axis, above the spectrometer entrance aperture. The area selected is a function of the aperture size demagnified back to the plane of the specimen. For example, if the image magnification is 100,000× at the recording plane and the *effective* entrance aperture size at the recording plane is 1 mm, then the area contributing to the spectrum is 10 nm. So, you might think that you can do high-spatial-resolution microanalysis without a probe-forming STEM. However, if you're analyzing electrons that have suffered a significant energy loss, they may have come from areas of the specimen well away from

the area you selected, because of chromatic aberration. This displacement d is given by

$$d = \theta\,\Delta f$$ [37.7]

where θ is the angle of scatter, typically <10 mrads, and Δf is the defocus error due to chromatic aberration given by

$$\Delta f = C_c \frac{E}{E_0}$$ [37.8]

where C_c is the chromatic aberration coefficient. So if we take a typical energy loss $\mathcal{E}$ of 284 eV (the energy required to eject a carbon K shell electron) and we have a beam energy of 100 keV, then the defocus due to chromatic aberration (with C_c = 3 mm) will be close to 10 μm, which gives an actual displacement, d, of 10^{-4} mm, or 100 nm. This figure is large compared to the value of 10 nm which we calculated without considering chromatic aberration effects.

> While TEM-image mode is good for gathering spectra with a large β and high-energy resolution, the price you pay is poorer spatial resolution.

In TEM diffraction mode, you select the area of the specimen contributing to the DP in the usual way. You can either use the SAD aperture, which has a lower limit of about 5 μm, or you can form a fine beam as in STEM, so that a CBED pattern appears on the screen. In the latter case, the area you select is a function of the beam size and the beam spreading, but is generally < 50 nm wide. Therefore, this method is best for high-spatial-resolution microanalysis; just as for XEDS microanalysis, STEM operating mode is recommended for EELS microanalysis.

- ■ Form an image in STEM mode.
- ■ Stop the probe from scanning.
- ■ Position it on the area to be analyzed.
- ■ Switch on the EELS; in a TEM/STEM you have to lift up the TEM viewing screen also!

37.5. WHAT YOU NEED TO KNOW ABOUT YOUR PEELS

As with XEDS, where there are several standard tests you need to perform to determine that all is well with the detector and the electronics, there are similar tests for the PEELS diode array and electronics. Some of these are described in the Gatan handbook and others have been proposed by Egerton *et al.* (1993). We'll discuss specific artifacts visible in the spectrum in the next chapter.

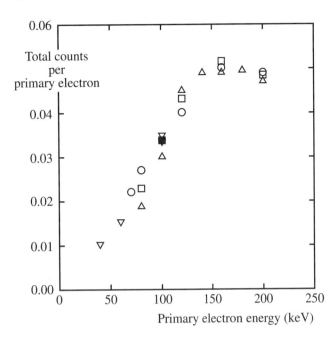

Figure 37.10. Nonlinear response of the diode array as a function of the beam energy. The response saturates at ~150 keV. Different symbols represent different dispersion settings.

Since increasing the keV means more electrons are generated in the scintillator, the sensitivity of the diode array should be linearly related to the electron energy. Egerton *et al.* (1993) have shown (see Figure 37.10) that, in fact, the Gatan diode-array response saturates at ~150 keV because of electron penetration. This nonlinearity doesn't affect quantification, since we typically make measurements over a very small energy range (<1 keV), but it means that there is no gain in count rate by operating >~150 keV. More important is the need for the YAG to respond linearly to different intensities incident on it; you should check that this is so by comparing the zero-loss intensity measured in a single 1-s readout with that recorded, say, in 40 readouts each of 0.025 s. In each case, you have to subtract the dark current (see Section 38.7). Obviously, the ratio of these two intensities should be unity for all levels of signal falling on the YAG. If it is not so, then you should consult the manufacturer.

37.6. IMAGING SPECTROMETERS

Two types of electron spectrometers are designed for energy-filtered imaging:

  ■ In-column spectrometers on Zeiss 902 and LEO 912 series TEMs for "electron-spectroscopic imaging" (ESI).

  ■ The Gatan Imaging Filter (GIF), which is a variation of the magnetic prism spectrometer.

Zeiss first used a mirror-prism system originally devised by Castaing and Henry (1962) and described by Zanchi *et al.* (1982). The drawback to the mirror-prism is the need to split the high-tension supply and raise the mirror to the same voltage as the gun. So LEO now use a magnetic omega (Ω) filter (Lanio *et al.* 1986). The Ω filter disperses the electrons in the column, as shown in Figure 37.11A. The spectrometer is placed in the TEM column between the intermediate and the projector lenses. Usually, you project an image into the prism, which is focused on a DP in the back focal plane of the intermediate lens. Therefore, the entrance aperture to the spectrometer selects an area of the specimen and the angle of collection is governed by the objective aperture (i.e., the same as image mode for the magnetic prism spectrometer). Electrons following a particular path through the spectrometer can be selected by the post-spectrometer slit. Thus only electrons of a given energy range, determined by the slit width, are used to form the image projected onto the TEM screen. ESI has several advantages over conventional TEM images, as we'll see in Section 40.3. We will also see then that the magnetic prism, which is primarily used for spectrometry, can also be used in a STEM to form energy-filtered images.

You can also change the microscope optics and project a DP into the prism, thus producing an energy-filtered DP on the TEM screen. Then, if you use the slit to select a portion of the DP, you get an energy-loss spectrum showing not only the intensity distribution as a function of energy but also the angular distribution of the electrons.

The GIF (Krivanek *et al.* 1992) shown in Figure 37.11B is basically a PEELS with an energy-selecting slit after the magnet and a two-dimensional slow scan CCD array detector rather than a single line of diodes. There are also more quadrupoles and sextupoles in the optics of the GIF. The first two quadrupoles before the slit increase the dispersion of the spectrometer onto the slit and the quadrupoles after the slit have two functions. Either they project an image of the spectrum at the slit onto the CCD, or they compensate for the energy dispersion of the magnet and project a magnified image of the specimen onto the CCD (which has advantages over the diode array in a conventional PEELS). In the first mode, the system is operating like a standard PEELS; in the second, it produces images (or DPs) containing electrons of a specific energy selected by the slit. Obviously, such a large number of variable sextupoles and quadrupoles could be a nightmare to operate without appropriate computer control, and this is built into the system. We'll describe energy filtering with the GIF in Section 40.3.

A

B

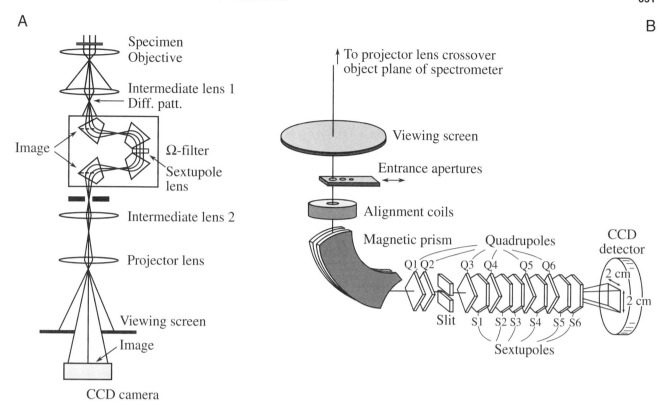

Figure 37.11. (A) Ray paths through the Ω filter system inserted in the imaging lens system of the LEO TEM. (B) The Gatan Imaging Filter attached to the TEM column after the imaging lenses, in the same position as a PEELS.

CHAPTER SUMMARY

We generally use a magnetic prism spectrometer for EELS. It is a simple device and very sensitive, but requires careful operation and an understanding of how it functions in combination with different TEM modes. PEELS is the preferred type of spectrometer, and it is best operated with the TEM in diffraction or STEM mode, or with a DSTEM. You have to know how to focus and calibrate it and how to determine the collection semi-angle, β. Once you can do this you're in a position to analyze energy-loss spectra, so in the next chapter we'll tell you what these spectra look like and what information they contain. If you have an Ω filter, or GIF, you can routinely form images or DPs with electrons of specific E.

REFERENCES

General References

Disko, M.M., Ahn, C.C., and Fultz, B., Eds. (1992) *Transmission Electron Energy Loss Spectrometry in Materials Science*, TMS, Warrendale, Pennsylvania.
Egerton, R.F. (1996) *Electron Energy-Loss Spectroscopy in the Electron Microscope,* 2nd edition, Plenum Press, New York.
Krivanek, O.L., Ed. (1991) *Microsc. Microanal. Microstruct.* **2** (2,3).
Krivanek, O.L., Ed. (1995a) *Microsc. Microanal. Microstruct.* **6** (1).
Krivanek, O.L., Ed. (1995b) *Ultramicroscopy* **59** (1–4).

Specific References

Castaing, R. and Henry, L. (1962) *C. R. Acad. Sci. Paris* **B255**, 76.
Egerton, R.F., Yang, Y.Y., and Cheng, S.Y. (1993) *Ultramicroscopy* **48**, 239.
Hillier, J. and Baker, R.F. (1944) *J. Appl. Phys.* **15**, 663.
Krivanek, O.L., Gubbens, A.J., Dellby, N., and Meyer, C.E. (1992) *Microsc. Microanal. Microstruct.* **3**, 187.
Lanio, S., Rose, H., and Krahl, D. (1986) *Optic* **73**, 56.
Metherell, A.J.F. (1971) *in Advances in Optical and Electron Microscopy,* **4** (Eds. R. Barer and V.E. Cosslett), p. 263, Academic Press, New York.
Zanchi, G., Kihn, Y., and Sevely, J. (1982) *Optik* **60**, 427.

The Energy-Loss Spectrum

38

CHAPTER PREVIEW

The term "energy-loss" spectrometry implies that we are only interested in inelastic interactions, but the spectrum will also contain electrons which have not lost any energy so we need to consider elastic scattering as well. We'll deal with three principal regions of the energy-loss spectrum:

- The zero-loss peak, which consists primarily of elastic forward-scattered electrons, but also contains electrons that have suffered minor (unresolvable) energy losses.
- The low-loss region up to an energy loss of ~50 eV contains electrons which have interacted with the weakly bound outer-shell electrons of the atoms in the specimen.
- Electrons in the high-loss region have interacted with the more tightly bound inner-shell or "core" electrons.

These different regimes of energy losses can give us different information about the specimen. The terminology is a bit vague but is generally accepted. The zero-loss peak defines the energy resolution and is essential in calibrating your spectrum. The electrons in the low-loss region have only interacted weakly with the atoms via their outer-shell electrons, so they contain information about the electronic properties of the specimen. The electrons in the high-loss region have "probed" the inner electron shells and therefore contain information characteristic of the atoms in the specimen.

653

We can also obtain information about how the atoms are bonded to one another, and even how the neighboring atoms are distributed around a specific atom. In principle, the energy-loss spectrum is far more useful than an XEDS spectrum. However, it is also far more complex. To understand its content you need a greater understanding of the physics of beam–specimen interactions. The spectrum also contains artifacts which we need to identify and minimize.

In this chapter we will discuss the different features of electron energy-loss spectra and go on to use these spectra in Chapters 39 and 40.

The Energy-Loss Spectrum

38

38.1. A FEW BASIC CONCEPTS

Back in Chapters 2–4 we talked about the difference between elastic and inelastic beam–specimen interactions and introduced the ideas of scattering cross sections and the associated mean-free path. It would be a good idea to remind yourself of those ideas before starting on this chapter. Briefly, you should recall that elastic scattering is an electron–nucleus interaction; the word "elastic" implies that there is no energy loss although a change in direction, and hence in momentum, usually occurs. Elastic scattering is usually manifest as Bragg diffraction in crystalline specimens. Inelastic scattering is primarily an electron–electron interaction and entails both a loss of energy and a change of momentum. Therefore, we have to be concerned with both the amount of energy lost and the direction of the electrons after they've come through the specimen. This latter point is one reason why the collection semiangle of the spectrometer is so important.

Remember, the cross section is a measure of the probability of a specific scattering event occurring and the mean free path is the average distance between particular interactions. Also, you must remember to distinguish between the definitions of scattering that will keep appearing.

Single scattering occurs when each electron undergoes at most one scattering event as it traverses the specimen.
Plural scattering (>1 scattering event) and multiple scattering (>20 scattering events) imply that the electron has undergone a combination of interactions.

We'll see that the energy-loss spectrum is most understandable when it represents single scattering. This ideal is approached when we have very thin specimens. In practice, most specimens are thicker than ideal and so we usually acquire plural-scattering spectra, and we may have to remove the plural-scattering effects. If multiple scattering occurs, the specimen is too thick for EELS and for much of TEM in general.

The principal inelastic interactions in order of increasing importance (and energy loss) are phonon excitations, inter- and intra-band transitions, plasmon excitations, and inner-shell ionizations. We've already introduced these processes back in Chapter 4 and we will emphasize inner-shell ionizations almost exclusively from here on. The two major characteristics of any inelastic scattering are the energy loss $\mathcal{E}$ and the scattering semiangle θ, and we summarize typical values in Table 38.1.

It's a little difficult to be specific about the values of the scattering angle because the angle varies with energy. In fact there are different definitions of scattering angle which you may come across, and these can be confusing. You can find derivations of the equations governing scattering in Egerton (1996).

The symbol θ in all cases refers to the scattering semiangle.

We will always assume that the scattering is symmetrical around the direct beam. The most important angle is θ_E, the so-called characteristic or most-probable scattering semiangle for an energy loss, $\mathcal{E}$. This angle is given by

$$\theta_E \approx \frac{\mathcal{E}}{2E_0} \qquad [38.1]$$

This equation is an approximation and it ignores relativistic effects, so you should only use it for rough calculations at and above 100 keV. We can be more precise and define θ_E as

**Table 38.1. Characteristics of the Principal
Energy-Loss Processes**

Process	Energy loss (eV)	θ_E (mrads)
Phonons	~0.02	5–15
Inter/intra-band transitions	5–25	5–10
Plasmons	~5–25	<~0.1
Inner-shell ionization	~10–1000	1–5

$$\theta_E \approx \frac{\mathcal{E}}{(\gamma m_0 v^2)} \qquad [38.2]$$

Here we have the usual definitions: m_0 is the rest mass of the electron, v is the electron velocity, and γ is given by

$$\gamma = \left(1 - \frac{v^2}{c^2}\right)^{-\frac{1}{2}} \qquad [38.3]$$

The electron velocity is v and c is the velocity of light. One other useful angle, θ_C, is the cut-off angle above which the scattered intensity is zero, and this is given by

$$\theta_C = (2\theta_E)^{\frac{1}{2}} \qquad [38.4]$$

In Table 38.1 we have given some typical values of θ_E. This is the scattering angle that we'll usually refer to from now on. Let's now move on to the energy-loss spectrum. We'll start at the low-energy end and proceed to higher-energy losses.

38.2. THE ZERO-LOSS PEAK

If your specimen is thin, the predominant feature in the energy-loss spectrum will be the zero-loss peak. As the name implies, this peak consists mainly of electrons that have completely retained the beam energy E_0. Such electrons may be forward scattered in a relatively narrow cone within a few mrads of the optic axis and constitute the 000 spot in the DP, i.e., the direct beam. If we were to tilt the incident beam so a diffracted beam entered the spectrometer, then it too would give a zero-loss peak. The scattering angles for diffraction ($2\theta_B$) are relatively large (~20 mrad) compared to the smaller collection angles in EELS, and so the diffracted beams rarely enter the spectrometer. Actually, we can also measure the intensity and energy of electrons as a function of their angular distribution, and we'll discuss this aspect briefly in Chapter 40.

Now the term "zero-loss peak" is really a misnomer for two reasons. First, our spectrometers have a finite energy resolution (at best ~0.3 eV) so the zero-loss peak will also contain electrons that have lost very small amounts of energy, mainly those that excited phonons. So in EELS in

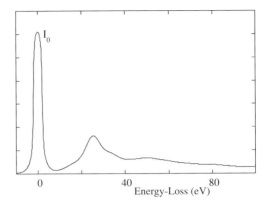

Figure 38.1. The intense zero-loss peak I_0 in a spectrum from stainless steel. The rest of the spectrum comprises energy-loss electrons which constitute a relatively small fraction of the total intensity in the spectrum.

the TEM we never resolve phonon losses. This is not a "great loss" since phonon-loss electrons don't carry any useful information anyway; they only cause the specimen to heat up. However, it does explain why we shouldn't really call this the zero-loss peak. Second, we can't produce a beam of monochromatic electrons; the beam has a finite energy range about the nominal value E_0. Despite this imprecision, we will continue to use the zero-loss terminology.

The zero-loss peak is usually a problem rather than a useful feature in the spectrum, because it is so intense that it can damage the scintillator or saturate the photodiode array. We don't collect it except under certain circumstances. Figure 38.1 shows the intense zero-loss peak in a spectrum. To the right of the peak is a relatively small peak, which is part of the low-loss spectrum. This small peak is where we start to get useful information, but you can also see immediately that the useful part of the spectrum is very much less intense than the somewhat useless zero-loss peak, and this is one of several fundamental problems in EELS.

38.3. THE LOW-LOSS SPECTRUM

We use the term "low-loss" to describe energy-loss electrons in the range up to about 50 eV. In this part of the spectrum we come across electrons that have set up plasmon oscillations or have generated inter- or intra-band transitions. Plasmons are by far the most important, so we'll look at these first.

38.3.A. Plasmons

Plasmons are longitudinal wave-like oscillations of weakly bound electrons. The oscillations are rapidly damped, typi-

Table 38.2. Plasmon Loss Data for 100-keV Electrons for Several Elements

Material	$\mathcal{E}_P$(calc) (eV)	$\mathcal{E}_P$ (expt) (eV)	θ_E (mrad)	θ_C (mrad)	λ_p(calc) (nm)
Li	8.0	7.1	0.039	5.3	233
Be	18.4	18.7	0.102	7.1	102
Al	15.8	15.0	0.082	7.7	119
Si	16.6	16.5	0.090	6.5	115
K	4.3	3.7	0.020	4.7	402

cally having a lifetime of about 10^{-15} s and so are quite localized to <10 nm. The plasmon peak is the second most dominant feature of the energy-loss spectrum after the zero-loss peak. The small peak beside the zero-loss peak in Figure 38.1 is a plasmon peak.

The energy $\mathcal{E}_P$ lost by the beam electron when it generates a plasmon of frequency ω_p is given by

$$\mathcal{E}_P = \frac{h}{2\pi}\omega_P = \frac{h}{2\pi}\left(\frac{ne^2}{\varepsilon_0 m}\right)^{\frac{1}{2}} \qquad [38.5]$$

where h is Planck's constant, e and m are the electron charge and mass, ε_0 is the permittivity of free space, and n is the free-electron density. Typical values of $\mathcal{E}_P$ are in the range 5–25 eV and a summary is given in Table 38.2.

> Plasmon losses dominate in materials with free-electron structures, such as Li, Na, Mg, and Al, but occur to a greater or lesser extent in all materials.

We even see a plasmon-like peak in spectra from materials with no free electrons (such as polymers) for reasons that are not well understood. From equation 38.5 you can see that $\mathcal{E}_P$ is affected by n, the free-electron density. Interestingly, n may change with the chemistry of the specimen. So in principle, measurement of the plasmon energy loss can give indirect microanalytical information, as we'll see later in Section 40.2. Plasmon-loss electrons also carry contrast information and therefore are important because they limit image resolution through chromatic aberration. We can remove them from the image by energy filtering, as we'll also describe in Section 40.3.

Because of the low values of λ_p, the characteristic scattering angles θ_E are very small, being typically <0.1 mrad (as listed in Table 38.2). So, plasmon-loss electrons are strongly forward-scattered. Their cut-off angle $\theta_C \sim 100\ \theta_E$. Hence if you use a β of only 10 mrad, you will gather virtually all the plasmon-loss electrons. Also, their line width $\Delta\mathcal{E}_P$ is at most a few eV.

A typical value of the plasmon mean-free path λ_p at AEM voltages is about 100 nm, and so it is reasonable to expect at least one strong plasmon peak in all but the thinnest specimens. Likewise, the number of individual losses should increase with the thickness of the specimen. Figure 38.2 shows the plasmon-loss spectra from thin and thick foils of pure Al. Since Al is a good approximation to a free-electron metal, the plasmon-loss process is the dominant energy-loss event. Plural plasmon scattering in thicker foils is a most important phenomenon because it eventually limits the interpretation of part of the spectrum containing chemical information from ionization losses in which we are really interested (see Section 38.4). The well-known properties of plasmon

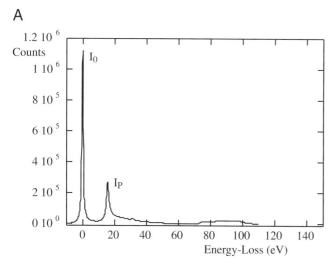
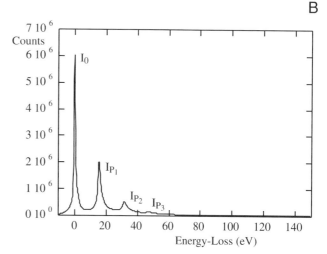

Figure 38.2. (A) The low-loss spectrum from a very thin sample of pure Al showing the intense zero-loss peak (I_0) and a small plasmon peak (I_p) at about 15 eV. (B) The low-loss spectrum from a thick specimen of pure Al showing several plasmon peaks.

loss electrons from several elements are summarized in Table 38.2.

The plasmon losses which we've just described all arise from interactions with the electrons in the interior of the specimen, but the incident electrons can also set up plasmon oscillations on the surface of the specimen. We can envisage these surface plasmons as transverse charge waves. Surface plasmons have about half the energy of bulk plasmons. Generally, however, the surface plasmon peak is much less intense than the volume plasmon peaks, even in the thinnest specimens.

38.3.B. Inter- and Intra-Band Transitions

An electron in the beam may transfer sufficient energy to a core electron to cause it to change its orbital state, for example, to a Bohr orbit of higher quantum number. We call these events "single electron interactions" and they result in energy losses of up to ~25 eV. Interactions with molecular orbitals such as the π orbitals produce characteristic peaks in this low-energy region of the spectrum, and it is possible sometimes to use the intensity variation in this part of the spectrum to identify a particular specimen. However, the details of the spectrum intensity variations due to single electron interactions are not well understood and cannot yet be predicted *a priori*.

Use of the low-loss spectrum for phase identification is only possible through a "fingerprinting" process by which the low-loss spectra of known specimens are stored in a library in the computer.

Spectra from unknown specimens may then be compared with the stored library standards. Figure 38.3 shows the low-loss spectra of Al and Al-containing compounds exhibiting differences in the detailed intensity variation. A collection of low-loss spectra from all the elements has been compiled in the EELS Atlas (Ahn and Krivanek 1983) and this can help with "fingerprinting" unknown specimens.

If the beam electron gives a weakly bound valence-band electron sufficient energy to escape the attractive field of the nucleus, then we've created a secondary electron (SE), of the sort used to give topographic images in the SEM and STEM. Typically, we give <20 eV to a SE and therefore the electrons causing SE emission appear in the same low-energy region of the spectrum as the inter- and intra-band transitions.

The weakly bound outer-shell electrons control the reaction of an atom to an external field and thus control the dielectric response of the material. We'll see in Chapter 40 that it is possible to get a measure of the dielectric constant by careful processing of the very low loss portion (<~10 eV) of the spectrum.

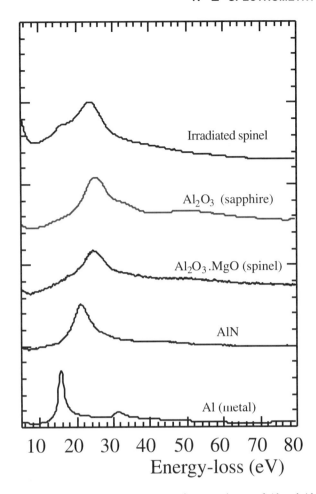

Figure 38.3. The low-loss spectrum from specimens of Al and Al-containing compounds, showing differences in intensity that arise from differences between the bonding in the different materials. The spectra are displaced vertically for ease of comparison.

38.4. THE HIGH-LOSS SPECTRUM

The high-loss portion of the spectrum above about 50 eV contains information from inelastic interactions with the inner or core shells.

38.4.A. Inner-Shell Ionization

When a beam electron transfers sufficient energy to a K, L, M, N, or O shell electron to move it outside the attractive field of the nucleus, as shown back in Figure 4.2, the atom is said to be ionized. As you know from the earlier chapters on X-ray analysis, the decay of the ionized atom back to its ground state may produce a characteristic X-ray, or an Auger electron. So the processes of inner-shell ionization-loss EELS and XEDS are different aspects of the same phenomenon. We are interested in ionization losses precisely because the process is characteristic of the atom involved and so the signal is a direct source of elemental information, just like the characteristic X-ray. We call the ionization-loss signal an "edge" for reasons we'll describe shortly.

> You should appreciate that detection of the beam electron that ionized the atom is independent of whether the atom emits an X-ray or an Auger electron. EELS is not affected by the fluorescence-yield limitation that restricts light-element X-ray analysis. This difference explains, in part, the complementary nature of XEDS and EELS.

Inner-shell ionization is generally a high-energy process. For example, the lightest solid element, Li, requires an input of ≥55 eV to eject a K-shell electron, and so the loss electrons are usually found in the "high-loss" region of the spectrum, above ~50 eV. K-shell electrons require much more energy for ejection as Z increases, because they are more strongly bound to the nucleus. The binding energy for electrons in the Uranium K shell is about 99 keV. So, as in XEDS, we tend to look for other lower-energy ionizations, such as the L and M edges, when dealing with high-Z atoms. Typically, we start to use the L edges when the K-shell energy exceeds ~1 keV (Na) and M edges when the L shell exceeds ~1 keV (Zn).

It's worth a short mention here about the nomenclature used for EELS edges. Just like in X-rays, where we have K, L, M, etc. peaks in the spectrum, we get ionization edges from K, L, M, etc. shell electrons. However, the greater energy resolution of the EELS spectrometer means that it is much easier to detect differences in spectra that arise from the presence of different energy states in the shell. For example:

■ The K-shell electron is in the 1s state and gives rise to a single K edge.
■ In the L shell, the electrons are in either 2s or 2p orbitals, and if a 2s electron is ejected, then we get an L_1 edge, and a 2p electron causes either an L_2 or L_3 edge.

The L_2 and L_3 edges may not be resolvable at lower ionization energies (e.g., they aren't in Al but they are in Ti), and sometimes we call this edge the $L_{2,3}$. The full range of possible edges is shown schematically in Figure 38.4, and you can see that other "dual" edges exist, such as the $M_{4,5}$. There will be more about this in Chapter 40.

Compared with plasmon excitation, which requires much less energy, the ionization cross sections are relatively small and the mean-free paths relatively large. As a result the ionization edge intensity in the spectrum is very much smaller than the plasmon peak, and becomes even smaller as the energy loss increases (look back to Figure 37.4A). This is another reason for staying with the lower-energy-loss (L and M) core edges. While the possibility of plural ionization events being triggered by the same electron is small in a typical thin foil, we'll see that the combination of an ionization loss with a plasmon loss is by no means uncommon, and this phenomenon distorts the resultant spectrum.

If you go back and look at Figure 4.2, you can see that a specific minimum-energy transfer from the beam electron to the inner-shell electron is required to overcome the binding energy of the electron to the nucleus and ionize the atom.

> This minimum energy constitutes the ionization threshold, or the critical ionization energy, E_C.

We define E_C as E_K for a particular K-shell electron, E_L for an L shell, etc. Of course, it is also possible to ionize an atom by the transfer of $\mathcal{E} > E_C$. However, the chances of ionization occurring become less with increasing energy above E_C, because the value of the cross section decreases with increasing energy. As a result, the ionization-loss electrons have an energy distribution that ideally shows a sharp rise to a maximum at E_C, followed by a slowly decreasing intensity above E_C back toward the background. This triangular shape is called an "edge."

> This idealized triangular or saw-tooth shape is only found in spectra from isolated hydrogen atoms, and is therefore called a hydrogenic ionization edge. Real ionization edges have shapes that approximate, more or less, to the hydrogenic edge.

You'll notice that this edge, shown in Figure 38.5A, has almost the same intensity profile as the "absorption edges" in X-ray spectroscopy. In reality, because we aren't dealing with isolated atoms but atoms integrated into a crystal lattice or amorphous structure, the spectra become more complex. The ionization edges are superimposed on a rapidly decreasing background intensity from electrons that have undergone random, plural inelastic scattering events (Figure 38.5B). The edge shape may also contain fine structure around E_C (Figure 38.5C) which is due to bonding effects, and is termed energy-loss near-edge structure (ELNES). More than ~50 eV after the edge, small intensity oscillations may be detectable (Figure 38.5D) due to diffraction effects from the atoms surrounding the ion-

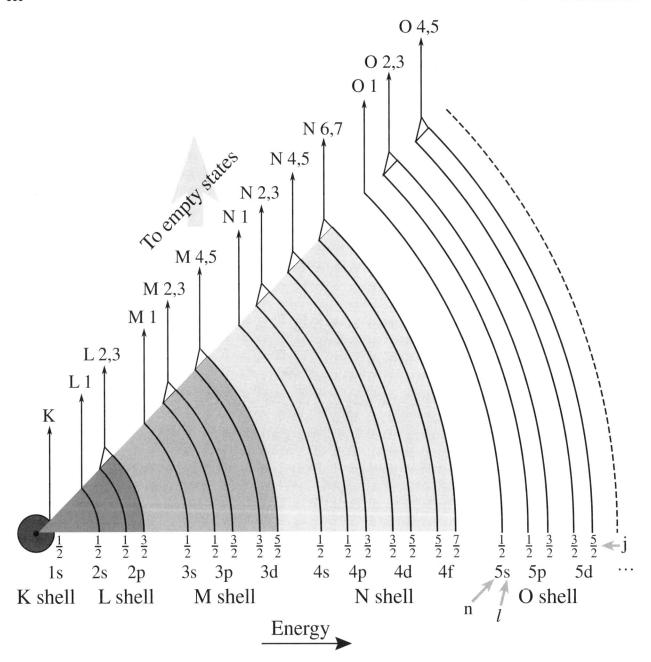

Figure 38.4. The full range of possible edges due to inner-shell ionization, and their associated nomenclature.

ized atom, and these oscillations are called extended energy-loss fine structure (EXELFS), which is analogous to extended X-ray absorption fine structure (EXAFS) in X-ray spectra, particularly those generated from intense synchrotron sources.

- ■ Fine structure before or around the peak is known as ELNES.
- ■ Small intensity oscillations >~50 eV after the edge due to diffraction effects are called EXELFS.

Finally, as we noted earlier, the ionization-loss electrons may also undergo further low-loss interactions. They may create plasmons, in which case the ionization edge contains plural scattering intensity ~15–25 eV above E_C, as shown in Figure 38.5E. So the resultant ionization edge is far more complicated than the simple Gaussian peak seen in an XEDS spectrum. Clearly, the edge details contain far more information about the specimen than a characteristic X-ray peak. From an X-ray spectrum you only get *elemental* identification rather than *chemical* in-

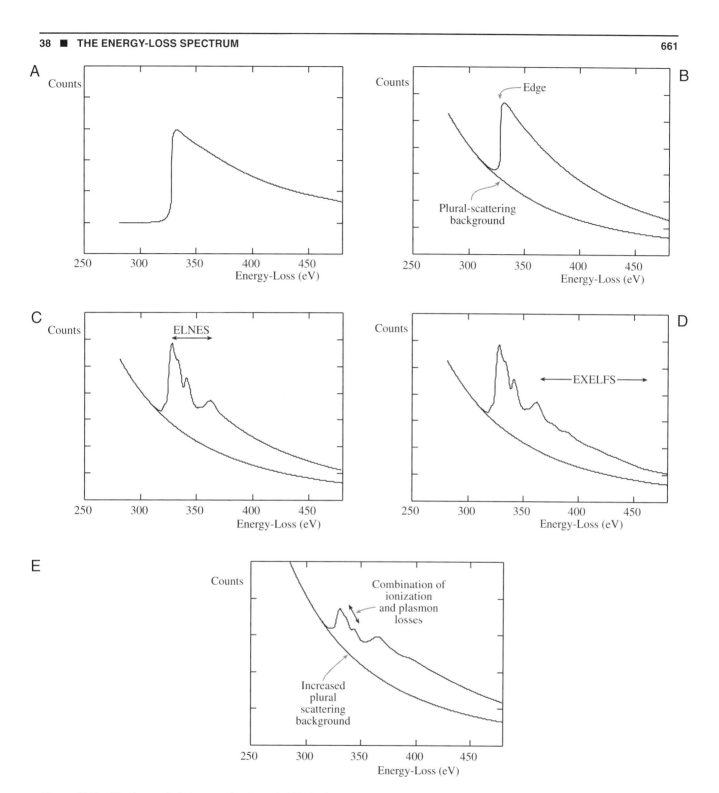

Figure 38.5. The characteristic features of an inner-shell ionization edge: (A) the idealized saw-tooth (hydrogenic) edge, (B) the edge superimposed on the background arising from plural inelastic scattering, (C) the presence of ELNES, (D) the EXELFS. (E) Plural scattering in a thick specimen, such as the combination of ionization and plasmon losses, distorts the post-edge structure and give an increase in the background level.

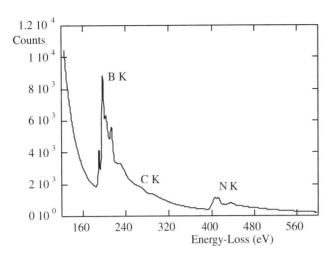

Figure 38.6. High-energy-loss spectrum from a particle of BN over a hole in a C film showing the B and N K-shell ionization edges superimposed on a rapidly decreasing background. A faint C K edge is also visible at ~280 eV.

formation, such as bonding, which is contained in the ELNES. Figure 38.6 shows a spectrum from BN on a C film. The various ionization edges show some of the features drawn schematically in Figure 38.5; we'll discuss these "fine structure" effects more in Section 40.1.

38.4.B. Ionization-Edge Characteristics

The angular distribution of ionization-loss electrons varies as $(\theta^2 + \theta_E^2)^{-1}$ and will be a maximum when $\theta = 0°$, in the forward-scattered direction. The distribution decreases to a half width at the characteristic scattering angle θ_E given by equation 38.1. This behavior is essentially the same as for plasmon scattering, but we have relatively large values of E_C compared to $\mathcal{E}_P$:

- θ_E ~5 mrad for ionization-loss electrons at $E_C = 1000$ eV, for a beam energy of 100 keV.
- The average plasmon-loss scattering was broadened to ~10–15 mrad.

The characteristic scattering angles for both plasmon and inner-shell ionization are still much lower than the characteristic scattering angles for phonon and elastic scattering, as we can appreciate from equation 38.1. The angular distribution varies depending on the energy loss, and because of the extended energy range of ionization-loss electrons above E_C, this can be quite complicated. For $\mathcal{E} \sim E_C$ the intensity drops rapidly to zero over about 10 mrad at θ_C, but as $\mathcal{E}$ increases above E_C the angular intensity distribution drops around $\theta = 0°$, but increases at larger scattering angles, giving rise to the so-called Bethe Ridge. Since this ef-

fect is irrelevant for most EELS studies, we'll ignore it, but you can find more information in Egerton (1996).

So, in the region immediately following the ionization edge the angular distribution of the electrons is generally confined to a semiangle of <10–15 mrad and drops to zero beyond this. In other words, like the plasmon-loss electrons, the ionization-loss electrons are very strongly forward-scattered. Consequently, efficient collection of most inelastically scattered electrons is a straightforward matter, since a spectrometer entrance aperture semiangle (β) of <20 mrad will collect the great majority of these electrons. As a result, collection efficiencies in the range 50–100% are not unreasonable, which contrasts with the situation in XEDS, where the isotropic generation of characteristic X-rays results in very inefficient collection. Figure 38.7 compares the collection of X-rays and energy-loss electrons in the AEM. Figure 38.8 shows the variation in collection efficiency for ionization-loss electrons as a function of both β and energy.

While the K edges in Figure 38.6 show reasonably sharp onsets, like an ideal hydrogenic edge, not all edges are similar in shape. Some edges have much broader onsets, spread over several eV or even tens of eV. The edge shape in general depends on the electronic structure of the atom but, unfortunately, we can't give a simple relationship between specific edge types and specific shapes. The situation is further complicated by the fact that the edge shapes change significantly depending on whether or not

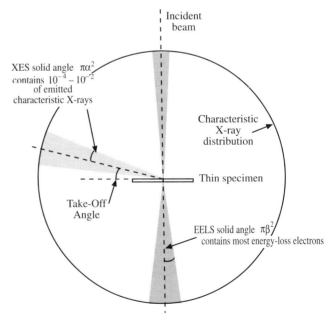

Figure 38.7. Comparison of the relative efficiencies of collection of EELS and XEDS. The forward-scattered energy-loss electrons are more efficiently collected than the uniformly emitted characteristic X-rays.

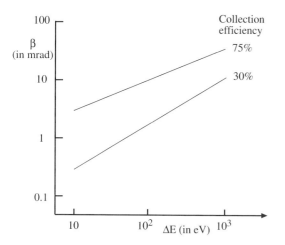

Figure 38.8. Variation in the collection efficiency of ionization-loss electrons as a function of their energy and the spectrometer collection semiangle, β.

certain energy states are filled or unfilled. For example, if you look below at Figure 38.9, the Ni L edge shows two sharp peaks, which are the L_3 and L_2 edges. (We'll discuss these details much more in Section 40.1.) These sharp lines arise because the ejected L shell electrons don't entirely escape from the atom and have a very high probability of ending up in unfilled d band states, which are present in Ni. In contrast, in Cu, in which the d band is full, the $L_{2,3}$ edge does not show these intense lines. Similar sharp lines appear in the $M_{4,5}$ edges in the rare earths. As if this were not enough, the details of the fine structure and edge shapes are also affected by bonding. For example, the Ni edge in NiO in Figure 38.9 is different from the Ni edge in pure Ni. To sort all this out it's best if you consult the EELS Atlas (Ahn and Krivanek 1983), which contains representative edges from all the elements and many oxides.

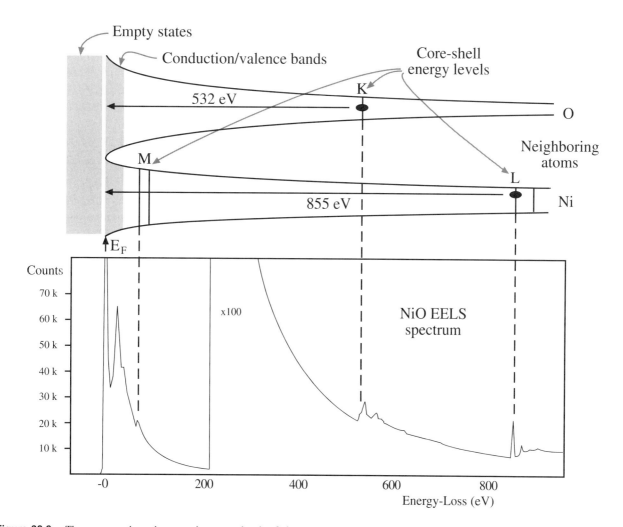

Figure 38.9. The correspondence between the energy levels of electrons surrounding adjacent Ni and O atoms and the energy-loss spectrum: the zero-loss peak is above the Fermi energy E_F, the plasmon peak is at the energy level of the conduction/valence bands, and the critical ionization energy required to eject specific K-, L-, and M-shell electrons is shown.

We can summarize the characteristics of the energy-loss spectrum by showing a complete spectrum from NiO containing both low- and high-loss electrons, as shown in Figure 38.9. In this figure we also compare the spectrum to the energy-level diagram for NiO. You can see that:

- The plasmon peak corresponds to the energy of the valence electron band just below the Fermi level (E_F).
- The relative energy levels of the ionized atom (K, L, or M) control the position of the ionization edge in the spectrum.
- The different density of states in the valence (3d) band of the Ni atom is indicated by shading at the top of the potential wells and is reflected in the characteristic, intense, near-edge, fine structure at the Ni L edge.

The electrons could also be given sufficient energy to travel into the conduction band well above E_F; as we just mentioned, in this case we see extended fine structure after the ionization edge. We'll discuss more details of such fine structure in the spectrum in Chapter 40.

Despite the very high collection efficiency of the spectrometer, the ionization edges, which are the major signal for elemental analysis, show relatively low intensity, have an extended energy range above the ionization energy, and ride on a rapidly varying, relatively high background. All these factors, as we shall see, combine to make quantitative microanalysis using EELS a difficult and less accurate technique when compared with XEDS. However, for the light elements the X-ray fluorescence yield drops to such low values, and absorption becomes so strong, even in thin specimens, that EELS is the preferred technique. Experimentally, the choice between the two is not simple, but below oxygen in the periodic table, EELS has shown better performance than XEDS and, for elements below boron, there is no sensible alternative to EELS for microanalysis at high spatial resolution.

38.5. ARTIFACTS IN THE SPECTRUM

The SEELS spectrum contains no artifacts of any consequence, unless it is grossly misaligned, in which case the beam may scatter through the slits or off the drift tube, giving distorted background intensities. These effects are easy to spot and correct.

Unfortunately, the highly efficient PEELS system generates more artifacts which you have to recognize and remove before analyzing the spectrum. Details are available in the Gatan manual, but here we'll summarize

the major problems (which are in addition to the point-spread function that we talked about in the previous chapter).

> All the individual diodes will differ slightly in their response to the incident electron beam, and therefore there will be a channel-to-channel gain variation in intensity. This will be characteristic of each individual diode array.

One way to determine the gain variation is to spread the beam uniformly over the array using at least the 3-mm entrance aperture and looking at the diode readouts, as shown in Figure 38.10. This is difficult with an FEG system because the probe is too small, and then it is necessary to scan the beam across the array, although this is not very satisfactory. Then you have to divide your experimental spectrum by this response spectrum to remove the gain variation. Alternatively, and this is recommended, you can gather two or more spectra with slight energy shifts (~1–2 eV) or spatial shifts between them and superimpose them electronically. The gain variation then disappears, as you can see if you look at Figure 38.12. Using a two-dimensional array, as in the GIF, removes this problem also.

Gathering many spectra and superimposing them can bring another problem, namely, that of readout noise. There are two kinds of readout noise, random and fixed. The random readout noise, or shot noise, arises from the electronics chain from the diode to the display, and is minimized by taking as few readouts as possible, and also by

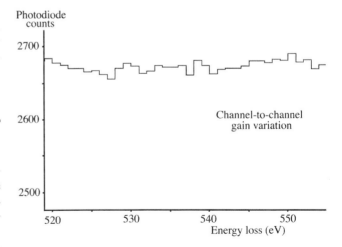

Figure 38.10. The variation in the response of individual diodes in the PEELS detection system to a constant incident electron intensity. A channel-to-channel gain variation is clear and each detector array has its own characteristic response function.

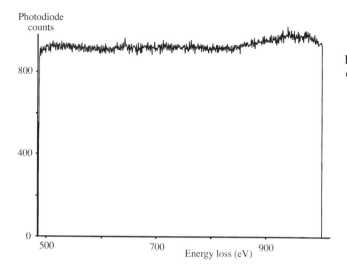

Figure 38.11. The intensity of the dark current which flows from the diode array when no electron beam is present.

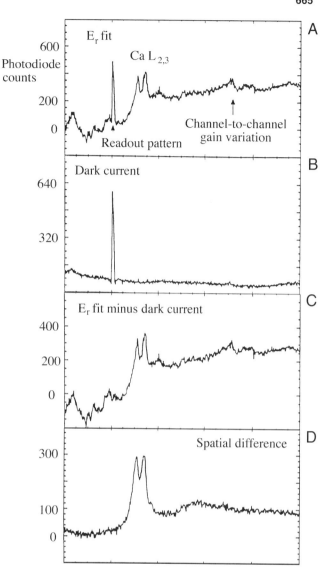

Figure 38.12. How to remove artifacts from a specimen: (A) A Ca L$_{2,3}$ edge spectrum showing both channel-to-channel gain variation and a faulty diode with a high leakage current which appears as a spike in the spectrum. The spike is referred to as the readout pattern and is present in every recorded spectrum. Subtracting the dark current (B) removes the spike (C) and a difference spectrum (D) removes the gain variation, leaving the desired edge spectrum.

cooling the diode array. Individual diodes may have high leakage currents which give a spike on the display. The fixed pattern readout noise is a function of the three-phase readout circuitry. All these effects will appear when there is no current falling on the diodes and together they constitute the dark current (see Figure 38.11). The dark current is small unless you have a bad diode array, and it is only a problem when there are very few counts in your spectrum or you have added together 10 or more spectra. Figure 38.12 shows some of these effects and how to remove them.

Finally, there is the problem of incomplete readout of the display. When the diodes are cooled, only ~95% of the signal is read out in the first integration, ~4.5% in the second, ~0.25% in the third, and so on. This is only a problem if you have saturated the diodes with an intense signal like the zero-loss peak. This peak then shows up as a ghost peak in the next readout and decays slowly over several readouts. So if a ghost peak appears, just run several readouts and it will disappear; this way you should never confuse a ghost peak with a genuine edge.

CHAPTER SUMMARY

The EELS spectrum varies in intensity over several orders of magnitude.

- The least useful signal (the zero-loss peak) is the most intense, and the most useful signals (the ionization edges) are among the least intense signals.
- The low-loss spectrum reflects beam interactions with loosely bound conduction and valence-band electrons.

Table 38.3. PEELS Artifacts and How to Eliminate Them

Noise name	Source	Elimination
Leakage current	Different diodes	Subtract dark current
Internal scanning noise	Electronics readout	Adjust the electronics and subtract the dark count
Nonuniform sensitivity	Diode sensitivities differ	Determine the response characteristic by sweeping the beam along the array and divide the real spectrum by this result, i.e., normalize the diodes

■ The high-loss spectrum contains small ionization edges riding on a strong plural-scattered background.

■ Differences in the energy onset of the ionization edges distinguish different elements in the specimen.

■ Differences in the fine structure of the edges reflect chemical (bonding) effects and structural (atomic arrangement) effects.

■ Artifacts can complicate spectrum interpretation, but they are well understood and easily removed.

We summarize the different sources of noise and how we eliminate this noise in Table 38.3.

REFERENCES

Specific References

Egerton, R.F. (1996) *Electron Energy Loss Spectroscopy in the Electron Microscope,* 2nd edition, Plenum Press, New York.

Ahn, C.C. and Krivanek, O.L. (1983) *EELS Atlas,* Gatan Inc., 780 Commonwealth Drive, Warrendale, Pennsylvania.

Microanalysis with Ionization-Loss Electrons

39

CHAPTER PREVIEW

In the previous two chapters we've described how to acquire an energy-loss spectrum and have also given you some idea of the information in such spectra. Most importantly, there are elemental composition data which can be extracted primarily from the high-loss ionization edges. In this chapter we'll examine how to get this information and quantify it. As we've already indicated, the prime use for these kind of data is light-element microanalysis, where EELS complements XEDS. First we'll remind you of the experimental variables over which you have control, because these are rather critical. Then we'll discuss how to obtain a spectrum and what it should look like for microanalysis. Next, we'll discuss the various quantification routines which, in principle, are just as straightforward as those for XEDS but in practice require a rather more sophisticated level of knowledge to carry them out successfully. Finally, we'll say a bit about spatial resolution and minimum detectability, although these topics aren't as important in EELS as they are in XEDS.

Microanalysis with Ionization-Loss Electrons

39

39.1. CHOICE OF OPERATING PARAMETERS

Perhaps a major reason why EELS is not as widespread as XEDS is the relative complexity of the experimental procedure and the number of variables which you have to define before you can get started. EELS is not yet a "turn-key" operation so you cannot simply place your specimen under the beam, switch on the spectrometer and the computer, and acquire a spectrum. This is in marked contrast to the situation in XEDS, where the high degree of software control means that the XEDS system is almost invariably ready to go when you push the "acquire" button. Furthermore, as you'll see, little useful information is present in the acquired EELS spectrum unless your specimen is very thin. Disko (1986) succinctly summarized the important experimental variables. We've already told you back in Chapter 37 how to control most of these factors. In this chapter, we'll go through all the parameters and indicate reasonable values for each.

- *Beam energy E_0:* It's probably best to use the highest E_0, unless doing so causes displacement damage or surface sputtering. A higher E_0 does reduce the scattering cross section and so you get reduced edge intensity. However, as E_0 increases, the plural-scattering background intensity falls faster than the edge intensity and so the ionization-edge signal-to-background increases and this is useful. The increase in signal-to-background varies with the particular edge but it is never a strong variation; so while we make a lukewarm recommendation to use the highest kV, it's not a good reason to justify purchasing an IVEM.
- *Convergence semiangle α:* You know how to control α with the C2 aperture and/or the C2 lens, but α is only important in the quantification process if it is larger than β. So if you op-

erate in TEM image or diffraction mode with a broad parallel beam, rather than STEM mode, you can ignore any effects of α; otherwise, use the correction factor we give in Section 39.7.
- *Beam size and current:* You control these factors by your choice of electron source, C1 lens, and C2 aperture. As usual, the beam size is important in limiting the spatial resolution in STEM mode, and the beam current controls the signal intensity. You have to make the same compromise between improved spatial resolution and loss of signal intensity, or vice versa, as we discussed at some length in Chapter 36 for XEDS.
- *Specimen thickness:* The specimen must be thin because then the plural-scattering contributions to the spectrum are minimized and quantification is most straightforward.

> Making your specimen as thin as possible is the most important part of EELS.

If your specimen is too thick, then you'll have to use deconvolution procedures to remove the effects of plural scattering. So we'll tell you how to determine the thickness from your spectrum and how to decide if you need to deconvolute the spectrum.
- *Collection semiangle β:* You know from Section 37.4 how to measure β in all operating modes. If you need lots of intensity and are happy with limited spatial resolution, use TEM-image mode with no objective aperture ($\beta \sim 100$ mrad). A small spectrometer entrance aperture would provide better energy resolution at the same time. If you want a small β to prevent contributions to the spectrum from high-angle scattering, use diffraction mode (TEM or STEM) and a small spectrometer entrance aper-

ture for good energy resolution. In the STEM case you also get good spatial resolution.

> Remember that a 5-mm entrance aperture gives β ~5 mrad at a camera length of ~800 mm.

Generally, for microanalysis β ~1–10 mrad is fine, so long as it's less than the Bragg angle for your particular specimen and orientation; but for EELS imaging, which we discuss in Section 40.3, 100 mrad may be necessary.

■ *Energy resolution:* ΔE is limited by your electron source, assuming you've focused the spectrum. In a SEELS, the slit width can control ΔE. Microanalysis and imaging do not require the best ΔE and ~5 eV will generally suffice. You really only need the best ΔE for ELNES, and plasmon-shift studies, both of which are somewhat esoteric pursuits. Use an FEG source and a PEELS if you want to do this kind of thing.

■ *Energy-loss range and spectrum dispersion:* The full spectrum extends out to the beam energy E_0, but the useful portion only extends to about 1 keV. Above this energy loss, the intensity is very low, and microanalysis by XEDS is both easier and more accurate, although arguably a little less sensitive. So you rarely need to collect a spectrum above about 1 keV and therefore, with a minimum of 1024 channels in the MCA display, 1 eV/channel is always a good starting point. You can easily select a higher display resolution if you want to look at a more limited region of the spectrum or if you want to see detail with $\Delta E < 1$ eV.

■ *Signal processing:* In SEELS, remember that your two choices (in which the spectral intensity is determined by the total current on the scintillator) are analog processing or single electron counting. You should collect the high-intensity, low-loss portion of the spectrum in analog mode and the lower-intensity, high-loss region in single-electron mode. The change in counting mode is most conveniently made at the same point in the collection process as the gain change. Set this point around 50–100 eV, well above the plasmon range but at lower E than most ionization edges. There is no equivalent of this variable in PEELS.

■ *Dwell time:* In SEELS, typical dwell times are in the range from 10 ms to 1 s per channel, depending on the number of channels in the spectrum and the intensity necessary to extract the analytical result. Because the magnetic prism is not very stable, it is unwise to collect spectra for periods longer than a few minutes. If more counts are required you should sum several spectra (see below), each recorded over a limited time range, with intermediate checks on the calibration. In PEELS, you set the dwell time (or integration time) such that the maximum intensity in the spectrum doesn't saturate the photodiode array, i.e., stay below 16,000 counts per acquisition in the most intense channel and sum as many spectra as you need to give sufficient counts for analysis.

■ *Number of sweeps:* It is better in both SEELS and PEELS to sum many spectra rather than gather one SEELS spectrum for several minutes, or saturate the PEELS detector. Remember also that each sweep in SEELS should be a reverse scan from high to low energy. Furthermore, if the intense zero loss has to be recorded, several minutes should elapse between each scan to ensure that the scintillator after-glow has subsided to below the normal dark-current output of the detector. In PEELS, multiple acquisitions can give rise to artifacts, as we discussed in Section 38.5.

So now you can see why EELS is not a straightforward turn-key operation. You must have a very good understanding of your TEM and spectrometer optics; be aware that the system is not very stable, needs recalibrating regularly and, PEELS, particularly, is prone to artifacts.

39.2. WHAT SHOULD YOUR SPECTRUM LOOK LIKE?

Before you analyze a particular spectrum, you should check three things:

■ Display the zero-loss peak to ensure that the spectrometer is giving you the necessary ΔE, if this is important.
■ Look at the low-loss portion of the spectrum; this gives you an idea of your specimen thickness.
■ Look for the expected ionization edges. If you can't see any edges, your specimen is probably too thick.

The first of these tasks is not critical, as we noted earlier. Regarding the second task, you'll see in Section 39.5 that, to a first approximation, if the plasmon peak intensity is less than about one-tenth the zero-loss peak, then the specimen is thin enough for microanalysis. Otherwise, you'll probably have to deconvolute plural-scattering effects from your experimental spectrum. For the third task, you should ideally see a discrete edge on a smoothly varying background, but you need to see at least a change in slope in the background intensity at any expected edge energies. The Gatan ELP program can identify and quantify such hidden peaks (see Section 1.5). If the background intensity is noisy, it will make quantification more difficult.

> An important parameter in determining the quality of your spectrum is the signal-to-background ratio which, in EELS, we call the jump ratio.

This is the ratio of the maximum edge intensity (I_{max}) to the minimum intensity (I_{min}) in the channel preceding the edge onset, as shown in Figure 39.1.

If the jump ratio is above ~5, for the carbon K edge at 284 eV from a standard (< 50 nm thick) carbon film at 100 kV, then your TEM-EELS system is operating satisfactorily. Figure 39.1 is a well-defined edge from a film of amorphous carbon. If you can't get such a jump ratio, then perhaps you need to realign the spectrometer, or find a thinner specimen. The jump ratio should increase with increasing kV.

39.3. QUALITATIVE MICROANALYSIS

As with XEDS, you should always carry out qualitative microanalysis to ensure that you have identified all the features in your spectrum. Then you can decide which edges to use for microanalysis.

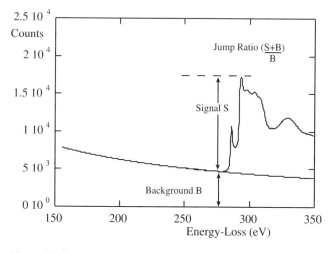

Figure 39.1. Definition of the jump ratio of an ionization edge which should be about 5–10 for the carbon K edge if the EELS is well aligned.

Qualitative microanalysis using ionization edges is very straightforward. Unlike XEDS, there are actually very few artifacts that can be mistaken for an edge. The most prominent artifact that may lead to misidentification is the so-called ghost edge from diode saturation in PEELS spectra (see Section 38.5). So long as you calibrate the spectrum to within 1–2 eV you can unambiguously identify the edge energy.

> We identify the ionization edge as the energy loss at which there is a discrete increase in the slope of the spectrum; this value is the edge onset, i.e., E_C, the critical ionization energy.

You have to be careful here: sometimes you'll see the edge energy defined somewhat arbitrarily half-way up the edge, e.g., at the π^* peak on the front of a C K edge. There is no strict convention, and very often L and M edges do not have sharp onsets anyhow.

Examination of a portion of a spectrum, such as that shown back in Figure 38.6, is usually sufficient to let you draw a definite conclusion about the identity of the specimen, which is BN on a C support film. In addition, it is wise to compare your spectrum with reference spectra from one of several EELS libraries that are available (Zaluzec 1981, Ahn and Krivanek 1983, Colliex 1984).

Remember that there are families of edges (K, L, M, etc.) just as there are families of peaks in X-ray spectra. As a rule of thumb, quantification is equally easy with K and L edges, but the accuracy of K-edge quantification is slightly better. Up to Z = 13 (Al) we usually use K edges, because any L edges occur at very low energy and are masked by the plasmon peak. Above Z = 13 you can use either K or L edges. Sometimes, there is the question of which edge is most visible. The K-edge onset is generally a bit sharper than the L edge, which consists of both the L_2 and L_3 edges and so may be somewhat broader. This is not always the case.

> L edges for Z = 19–28 and 37–45 are characterized by intense near-edge structure, called white lines. M edges for Z = 55–69 have similar intense lines.

These white lines, which we first saw back in Figure 38.9, are so named because of their appearance in early, photographically recorded energy-loss spectra; more details are given in Section 40.1. If you have to use the M, N, or O edges without any white lines, you should know that they are very broad, with an ill-defined threshold, and quantification is only possible with standards, as we'll see shortly.

The energy-loss spectrum clearly does not lend itself to a quick "semiquantitative" analysis, as we can do with XEDS. For example, the spectrum in Figure 38.6 comes from equal numbers of B and N atoms, but the intensities in the B and N edges are markedly different. This difference arises because of the variation in ionization cross section with $\mathcal{E}$, the strongly varying nature of the plural-scattering background, and the edge shape, which causes the C and N K edges to ride on the tails of the preceding edge(s).

Example
Sometimes, qualitative analysis is all that you need to do. Figure 39.2A and Figure 39.2B show images and spectra from two small precipitates in an alloy steel. The spectra show a Ti L_{23} edge in both cases, and C and N K edges in Figure 39.2A and Figure 39.2B, respectively. It does not take much effort to deduce that the

first particle is TiC because it is the only known carbide of Ti, but the nitride could be either TiN or Ti_3N. To determine which of the two it is, you have to carry out full quantification, which we'll discuss shortly. You should note that such clear discrimination between TiC and TiN in Figure 39.2 would be difficult using windowless XEDS, because the energy resolution is close to the separation of the Ti L (452 eV) and the N K (392 eV) X-ray peaks. In addition, the DPs from both phases are almost identical, so this problem is a perfect one for EELS.

39.4. QUANTITATIVE MICROANALYSIS

To quantify the spectrum, you have to extract intensity in the ionization edge(s) by removing the plural-scattering background and integrating the intensity (I) in the edge.

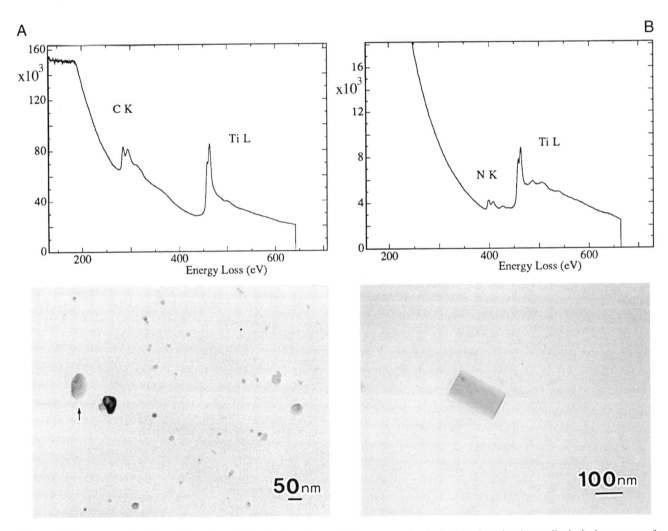

Figure 39.2.   Images of small precipitates in a stainless steel specimen, and the corresponding ionization edges showing qualitatively the presence of Ti, C, and N. Thus the precipitates can be identified as (A) TiC and (B) TiN, respectively.

Then you have to determine the sensitivity factor, that is, you need to determine the number of atoms N responsible for I. This sensitivity factor is called the "partial ionization cross section." We'll see that it plays a similar role to the k_{AB} factor in X-ray microanalysis. If you go back and look at Figure 38.5, you'll see how an ionization edge is built up from several contributions. The process of quantification in essence involves stripping away (or ignoring) the various contributions until you're left with Figure 38.5A, which contains the single-scattering "hydrogenic" edge intensity.

39.4.A. Derivation of the Equations for Quantification

The equations we use for quantitative analysis have been derived, refined, and applied by Egerton and co-workers. The following derivation is a summary of the full treatment by Egerton (1996), which itself draws on work carried out over the preceding two decades.

We'll assume that we are quantifying a K edge, although the basic approach can be used for all edges. The K-shell intensity above background, I_K, is related to the probability of ionization, P_K, and the total transmitted intensity, I_T

$$I_K = P_K I_T \qquad [39.1]$$

This equation assumes that the intensities are measured over the complete angular range (0–4π sr), which of course is not the case, but we'll correct for this later. In a good thin specimen we can approximate I_T to the incident intensity, neglecting backscatter and absorption effects. Now, this is the important point:

> If we assume also that the electrons contributing to the edge have only undergone a single ionization event, then we can easily obtain an expression for P_K

$$P_K = N\sigma_K \exp\left(\frac{t}{\lambda_K}\right) \qquad [39.2]$$

where N is the number of atoms *per unit area* of the specimen (thickness t) that contribute to the K edge. The assumption of a single K-shell ionization event with cross-section σ_K is reasonable, given the large mean free path (λ_K) for ionization losses; but it explains why you have to make thin specimens. It also means that the exponential term is very close to unity, and so

$$I_K \approx N\sigma_K I_T \qquad [39.3]$$

and therefore

$$N = \frac{I_K}{\sigma_K I_T} \qquad [39.4]$$

Thus we can measure the absolute number of atoms per unit area of the specimen simply by measuring the intensity above background in the K edge and dividing it by the total intensity in the spectrum and the ionization cross section. We can easily extend this expression to a spectrum containing two edges from elements A and B, in which case the total intensity drops out and we can write

$$\frac{N_A}{N_B} = \frac{I_K^A \sigma_K^B}{I_K^B \sigma_K^A} \qquad [39.5]$$

Similar expressions apply to L, M edges, etc., and combinations of edges can be used. So you see that if you are quantifying more than one element then you don't need to gather the zero-loss peak, and this makes life much easier for the spectrometer scintillator or diode array.

In both equations 39.4 and 39.5 we assumed that we could accurately subtract the background under the ionization edge and that we know σ. Unfortunately, as you'll see, both background subtraction and determination of σ are nontrivial and limit the accuracy of quantification. We will discuss these points later, but initially we must take account of the practical realities of spectrum acquisition and modify the equations accordingly.

First, you can't gather the whole of the energy-loss spectrum out to the beam energy, E_0, because above 1–2 keV the intensity decreases to a level close to the system noise. Furthermore, while ionization-loss electrons can theoretically have any energy between E_C and E_0, in practice the intensity in the edge falls to the background level within about 100 eV of the ionization threshold, E_C. In addition, the background extrapolation process becomes increasingly inaccurate beyond about 100 eV, and so it is imperative to restrict the integration of spectral intensities to some window, Δ, usually in the range of 20–100 eV. So we modify equation 39.4 to give

$$I_K(\Delta) = N\sigma_K(\Delta)I_T(\Delta) \qquad [39.6]$$

The term $I_T(\Delta)$ is more correctly written as $I_\ell(\Delta)$, where I_ℓ is the intensity of the zero-loss (direct beam) electrons combined with the low-loss electrons over an energy loss window Δ. Only if we have true single scattering can we use I_T, and we'll discuss the conditions for this later.

As we discussed, EELS has the tremendous advantage that the energy-loss electrons are predominantly forward-scattered and so you can easily gather most of the signal. As a result, the technique is inherently far more efficient than XEDS. However, because we cannot physically collect the spectrum over 4π sr, but are limited by our choice of collection semiangle β, we must further modify the equation and write

$$I_{\mathrm{K}}(\beta\Delta) = N\sigma_{\mathrm{K}}(\beta\Delta)I_\ell(\beta\Delta) \qquad [39.7]$$

The factor $\sigma_{\mathrm{K}}(\beta\Delta)$ is termed the "partial ionization cross section," from this equation, therefore, the absolute quantification for N is given by

$$N = \frac{I_{\mathrm{K}}(\beta\Delta)}{I_\ell(\beta\Delta)\sigma_{\mathrm{K}}(\beta\Delta)} \qquad [39.8]$$

For a ratio of two elements A and B, the low-loss intensity drops out

$$\frac{N_{\mathrm{A}}}{N_{\mathrm{B}}} = \frac{I_{\mathrm{K}}^A(\beta\Delta)\sigma_{\mathrm{K}}^B(\beta\Delta)}{I_{\mathrm{K}}^B(\beta\Delta)\sigma_{\mathrm{K}}^A(\beta\Delta)} \qquad [39.9]$$

We can draw a direct analogy between this equation and the Cliff–Lorimer expression (equation 35.2) used in thin-foil XEDS. In both cases, the composition ratio C_A/C_B or N_A/N_B is related to the intensity ratio I_A/I_B through a sensitivity factor, which we call the k_{AB} factor in XEDS and which in electron spectrometry is the ratio of two partial cross sections, σ^B/σ^A.

Remember that the major assumption in this whole approach is that the electrons undergo *a single scattering event*. In practice, it's difficult to avoid plural scattering, although in very thin specimens the approximation remains valid, if errors of ±10–20% are acceptable. If plural scattering is significant then the spectrum must be deconvoluted, and we will discuss ways to do this in Section 39.6 when we describe the limitations of specimen thickness. You should also note when using the ratio equation that your analysis is a lot better if the two edges are similar in shape, i.e., both K edges or both L edges, otherwise the approximations inherent in equation 39.9 will be less accurate.

In summary then, these equations give us an absolute value of the atomic content of the specimen or a ratio of the amounts of two elements. You have to carry out two essential practical steps:

- The background subtraction to obtain I_{K}.
- The determination of the partial ionization cross section $\sigma_{\mathrm{K}}(\beta\Delta)$.

So now you can see why it is important to know β.

39.4.B. Background Subtraction

The background intensity comes from plural-scattering events which are usually associated with outer-shell interactions. In the spectrum the background appears as a rapidly changing continuum decreasing from a maximum just after the plasmon peak at about 15–25 eV, down to a minimum at which it is indistinguishable from the instrumental noise, typically when $\mathcal{E} \sim$ 1–2 keV. In addition to the true plural scattering, there is also the possibility of single-scattering contributions to the background from the tails of preceding ionization edges. Because of the complexity of the many combinations of plural-scattering processes, it has not proven possible to model the background from first principles to the same degree that is possible in XEDS using variations on Kramers' Law. There are three ways commonly used to remove the background:

- Curve fitting.
- The graphical method.
- Using difference spectra.

We'll now describe these in some detail.

Curve Fitting: You select a window δ in the background before the edge onset and fit a curve to the channels. Then you extrapolate the curve over the desired energy window Δ under the edge. This process is shown schematically in Figure 39.3, and experimentally in Figure 39.4.

We assume that the energy dependence of the background has the form

$$I = A\,\mathcal{E}^{-r} \qquad [39.10]$$

where I is the intensity in the channel of energy loss $\mathcal{E}$, and A and r are constants for a particular curve fit. The fitting parameters are only valid over a limited energy range because they depend on $\mathcal{E}$. The exponent r is typically in the

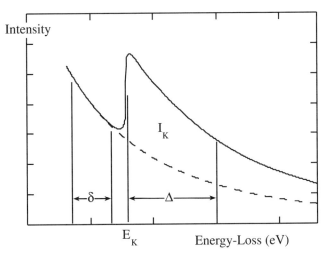

Figure 39.3. The parameters required for background extrapolation and subtraction under an ionization edge. The pre-edge fitting window δ is extrapolated over a post-edge window Δ then subtracted to give the edge intensity I_{K}.

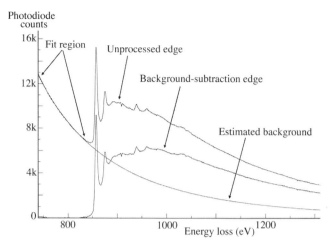

Figure 39.4. A Ni L$_{2,3}$ edge before and after background subtraction. The fit region before the unprocessed edge is extrapolated to give the estimated background, which is then removed, leaving the background-subtracted edge.

range 2–5, but *A* can vary tremendously. We can see some trends in how *r* varies. The value of r decreases as:

■ The specimen thickness, *t*, increases.

■ The collection semiangle, β, increases.

■ The electron energy loss, $\mathcal{E}$, increases.

The fit to the tail of a preceding edge also shows a similar power-law dependence on the plural-scattering background, and may be fitted in a similar manner, i.e., $I = B\mathcal{E}^{-s}$. The energy range δ over which you fit the background should not be <~10 channels, and at most not >~30% of E_K. In practice, however, you might not be able to fit the background over such a wide window if another edge is present within that range.

You should choose the extrapolation window, Δ, such that the ratio of the finish to the start energies, $\mathcal{E}$(finish)/$\mathcal{E}$(start), is <1.5. In Figure 39.4, the extrapolation window (~450 eV) is a little larger than ideal. So, Δ is smaller for lower edge energies. Using larger windows, although improving the statistics of the total edge integration, eventually reduces the accuracy of the final quantification because the fitting parameters *A* and *r* are only valid over ~100 eV. If there's a lot of near-edge structure, either use a larger Δ to minimize its effect or avoid it in the extrapolation window, unless the quantification routine can handle it.

Instead of the simple power-law fit, you can use any expression such as an exponential, polynomial, or log-polynomial, so long as it provides a good fit to the background and gives acceptable answers for known specimens. Polynomial expressions can behave erratically if you extrapolate them over a large Δ, so use them cautiously. Generally, the power law seems adequate for most

purposes except close to the plasmon peaks ($\mathcal{E}$ <~100 eV). Clearly, the background channels closest to the edge onset will influence the extrapolation most strongly, and various weighting schemes have been proposed. A noisy spectrum will be particularly susceptible to poor fitting, unless some type of weighting is used.

We can judge the "goodness of fit" of a particular power-law expression qualitatively by looking at the extrapolation to ensure that it is heading toward the post-edge background and not substantially under- or over-cutting the spectrum. More quantitatively, we can assign a χ^2, chi-squared, value based on a linear least-squares fit to the experimental spectrum. The least-squares fit can be conveniently tied in with a weighting scheme using the expression

$$\chi^2 = \sum_i \frac{(y - y_i)^2}{y^2} \qquad [39.11]$$

where y_i is the number of counts in the *i*th channel and $y = \ln_e I$. The squared term in the denominator ensures suitable weighting of the channels close to the edge.

Difference Spectra: You can also remove the background using a first-difference approach (which is equivalent to differentiating the spectra). This method is particularly suited to PEELS since it simply involves taking two spectra, offset in energy by a few eV, and subtracting one from the other. As shown in Figure 39.5, the difference process results in the slowly varying intensity (i.e., background) being reduced to zero and the rapidly varying intensity features (i.e., ionization edges) showing up as classical difference-peaks, similar to what you may have seen in an Auger spectrum. This is the *only* way to remove the background if your specimen thickness changes over the area of analysis, and it also has the advantage that it sup-

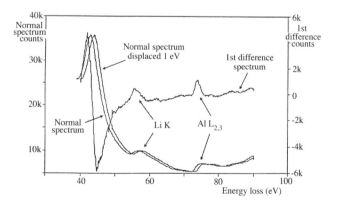

Figure 39.5. First-difference method of background subtraction, showing two PEELS spectra from a specimen of Al-Li displaced by 1 eV and subtracted to give a spectrum in which the background intensity falls to zero and the small Li K and Al L$_{2,3}$ edges are clearly revealed.

presses spectral artifacts common to PEELS, particularly the channel-to-channel gain variation.

Another kind of difference method involves convoluting the experimental spectrum with a filter function. A top-hat filter function, similar to the one we described in Section 35.2 for background subtraction in XEDS, gives a second-difference spectrum which also removes the background but exacerbates some artifacts.

39.4.C. Edge Integration

The edge integration procedure you use depends on how you removed the background. If you used a power-law approach, then remember that there is a limit over which the edge integration window Δ is valid. The value of Δ should be large enough to maximize the integrated intensity, but not so large that the errors in your background extrapolation dominate. Often, the presence of another edge limits the upper end of the integration window. The lower end is usually defined from the edge onset, E_K, but if there is strong near-edge structure, such as in the B K edge or the Ca L_{23} edge, then your integration window should start at an energy above these, unless the quantification schemes can handle fine structure effects (see below). If you subtracted the background using a first-difference approach, then you determine the peak intensity by fitting the experimental spectrum to a reference spectrum from a known standard using multiple least-squares fitting. We'll talk more about this when we discuss deconvolution of spectra.

39.4.D. The Zero-Loss Integral

Remember from equation 39.8 that if you want *absolute* quantification of N, then you have to integrate the low-loss spectrum I_ℓ out to about 50 eV. In a SEELS you should always do this using a reverse scan to avoid any problems with after-glow of the scintillator, and in PEELS you must be careful to integrate for a short enough time so you don't saturate the diode array. If you are doing a ratio, then I_ℓ is not needed (equation 39.9).

39.4.E. The Partial Ionization Cross Section

There are several ways we can determine the partial ionization cross section, $\sigma(\beta\Delta)$, which is the sensitivity factor relating intensity (I) to the number of atoms (N). We either use a theoretical approach or compare the experimental spectra with known standard spectra.

Theoretical Calculation: The most common approach is that due to Egerton (1979, 1981), who produced two short computer programs to model the K- and L-shell partial cross sections. The programs are called SIGMAK

and SIGMAL, respectively. They are public domain software and are available in Gatan's ELP software, but the code is also given in Egerton's book.

> The cross sections are modeled by approximating the atom in question to an isolated hydrogen atom with a charge on the nucleus equal to the atomic number Z of the atom, but with no outer-shell electrons.

While at first sight this is an absurd approximation, the approach is tractable because the hydrogen-atom wave function can be expressed analytically by Schrödinger's wave equation, which can be modified to account for the increased charge. Because this treatment neglects the outer-shell electrons, it is best suited to K-shell electrons, and Figure 39.6 shows a comparison between the measured N K-shell intensity and that computed using SIGMAK. As you can see, the SIGMAK hydrogenic model essentially ignores the near-edge and post-edge fine structure (which would be absent in the spectrum from a hydrogen atom), but still gives a very good fit, on average, to the experimental edge. Figure 39.7 compares the Cr L edge with the SIGMAL model. The L-shell fit is almost as good as the K fit, but the white lines are imperfectly modeled. These programs are very widely used since they are simple to understand and easy and quick to apply.

There is another theoretical approach which uses empirical parameterized equations to calculate the terms that modify σ for the effects of β and Δ. Both Joy (1986b) and Egerton (1989) have given relatively simple expressions, amenable to evaluation on a hand-held calculator, which you can look up if you wish. Joy's parameterization approach and the SIGMAK/SIGMAL models give good agreement, as shown in Figure 39.8. There are more com-

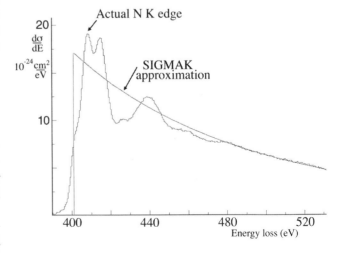

Figure 39.6. Comparison of an experimental N K edge and the hydrogenic fit to the edge using the SIGMAK program.

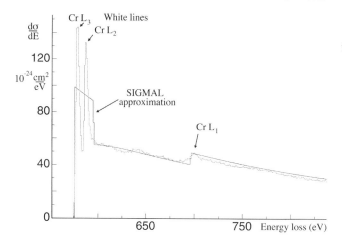

Figure 39.7. Comparison between an experimental Cr $L_{2,3}$ edge and a modified hydrogenic approximation to the edge obtained using the SIGMAL program. The fit makes no attempt to model the intense white lines, but only makes a rough estimate of their average intensity.

plex methods available which calculate the cross section in a more realistic way than the hydrogenic model, e.g., using Hartree–Slater models or atomic-physics approaches, which are better for the more complex L and M edges (Rez 1989). Egerton (1993) has compared experimental and theoretical cross sections, and the M-shell data (which are the worst case) are shown in Figure 39.9. The data are actually plotted in terms of the oscillator strength f (which is a measure of the response of the atom to the incident electron). This term is the integral of the generalized oscillator strength, which is proportional to the differential cross section, so just think of f as proportional to σ. There is still relatively poor agreement between experiment and theory for the M shell, as well as between the atomic and hydrogenic theoretical models. Similar data in Egerton's paper show

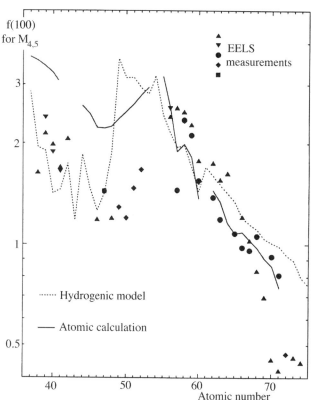

Figure 39.9. Comparison of the experimental and theoretical approaches to determination of the M-shell ionization cross section shown in terms of the variation in the dipole oscillator strength (f) as a function of atomic number. The data points are different experimental measurements, the solid line is a fundamental atomic calculation, and the dotted line is a hydrogenic calculation.

better agreement for K and L shells. These models, while more precise, require substantially longer computing time, but this is fast becoming less of a problem. Given the other sources of error in EELS microanalysis, you rarely need to go to such lengths to obtain a better value of $\sigma(\beta\Delta)$, and you should generally stick with the SIGMAK/L approach for routine quantification.

Experimental Determination: Rather than calculating σ theoretically, you can generate a value experimentally using known standards. This approach is, of course, exactly analogous to the experimental k-factor approach for XEDS quantification in which the cross section is automatically included (along with the fluorescence yield and other factors). It is surprising at first sight that the classical XEDS approach of using standards has not been widely used in EELS, but the reason is obvious when you remember the large number of variables that affect the EELS data. The standard and unknown must have the same thickness and the same bonding characteristic, and the spectra must be gathered under identical conditions; in particular β, Δ, E_0, and t must be the same.

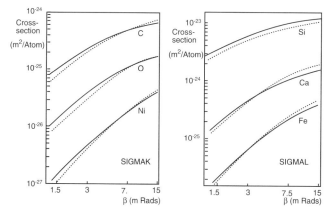

Figure 39.8. Comparison of the SIGMAK and SIGMAL hydrogenic models (full lines) for the ionization cross section with the parametric model (dotted lines).

> Again, it is the problem of thickness measurement that appears to be the main limitation to improving the accuracy of microanalysis.

Despite these limitations, data are available comparing cross sections of various elements, relative to oxygen (Hofer *et al.* 1988), just as *k* factors are determined relative to Si or Fe. Malis and co-workers (Malis *et al.* 1987, Malis and Titchmarsh 1988) have also produced a large number of experimental cross sections for light-element compounds. It is intriguing to note that the experimental approach appears increasingly popular in EELS, while in XEDS the reverse trend, toward more theoretical modeling of *k* factors, seems to be the case!

The SIGMAK/L programs may introduce large errors when quantifying the lightest metallic elements, Li and Be, so for the most accurate quantification of these elements the standards approach is still the best.

Example

In a study of Al-Li alloys (Liu and Williams 1989), a homogeneous sample of Al-12.7 at.% Li was used as a standard; *t* was determined from the relative intensities of the first plasmon peak and the zero-loss peak. The integrated intensity ratio for the Li K/Al L_{23} edges was determined after background subtraction to be 0.106±0.006. This number was the average of six separate spectra, and the errors were based on a student *t* analysis at the 95% confidence limit. From equation 39.9, the Li/Al partial-cross-section ratio was calculated to be 1.37±0.07. An Al-Li specimen containing an intermetallic of unknown Li content was then examined and 13 spectra obtained which gave an average Li K/Al L_{23} intensity ratio of 0.188±0.009. Combining this ratio with the partial ionization cross section and substituting back into equation 39.9 gives the composition of the intermetallic as Al-20.5±1 at.%Li. This result and others are given in Figure 39.10, which shows the low-Li portion of the Al-Li phase diagram, determined through direct Li composition measurements. For comparison, the partial ionization cross section was also determined from the SIGMAK/L programs, and the ratio was 0.969, ~30% less than that obtained using the standard. While this is a large difference compared to most SIGMAK/L calculations, it still sounds a note of caution against unquestioning use of the calculated cross sections.

In summary, there are two approaches to the determination of $\sigma(\beta\Delta)$: theoretical calculation and experimental measurement. In contrast to XEDS, the theoretical approaches dominate. There is good evidence that, particularly for the lighter elements for which EELS is best suited, the simple and quick hydrogenic model is usually ade-

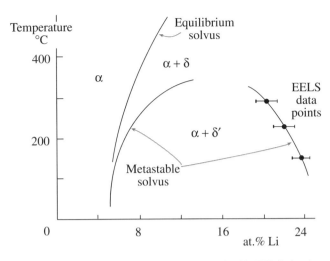

Figure 39.10. The Al-Li phase diagram determined by EELS, showing a variation in the Li content of the metastable Al_3Li (δ') phase as the temperature is raised. The equilibrium phases are α (Al-Li) solid solution and δ (Al-Li) intermetallic.

quate. However, for the heavier elements, where the M shell is used for analysis, tedious experimental data are still the best option. Of course, for such elements it is probably better to revert to XEDS analysis anyhow.

So now we're in a position where we have all the data needed to solve the quantification equations. However, our assumption all along has been that the spectra were the result of single scattering and we neglected plural scattering. Now in practice there will *always* be some plural-scattering contribution to the ionization edges.

> The combination of a plasmon interaction and an ionization will show up as a bump about 15–25 eV past the onset of the edge.

This effect is shown schematically back in Figure 38.5F and, if you look ahead, in Figure 39.15. So how do we go about correcting for this? We can either make our specimens so thin that plural scattering is negligible, or we can deconvolute the spectra. The former approach is possible, but you have to be lucky or exceptionally skilled at specimen thinning. The latter approach is mathematically simple, but can be misleading if not done properly, so we will need to examine deconvolution in more detail; but let's look at how we determine *t* because EELS offers us a simple method for this.

39.5. MEASURING THICKNESS FROM THE ENERGY-LOSS SPECTRUM

There is thickness information in the energy-loss spectrum since the amount of all inelastic scatter increases with

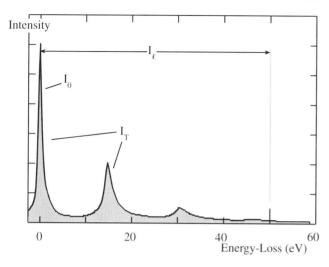

Figure 39.11. Definition of the zero-loss intensity I_0, the total intensity I_T, and the low-loss (I_ℓ) intensity required for thickness determination.

specimen thickness. In principle we have to measure the intensity under the zero-loss peak (I_0) and ratio this to the total intensity in the spectrum (I_T), as defined in Figure 39.11. But in practice the intensity in the EELS spectrum falls so rapidly with increasing energy loss that we can reasonably approximate I_T to the intensity in the low-loss portion of the spectrum I_ℓ, out to about 50 eV. The relative intensity of the zero loss and the total intensity is governed by the average mean free path (λ) for energy losses up to 50 eV, and thus

$$t = \lambda \ln\left(\frac{I_\ell}{I_0}\right) \qquad [39.12]$$

All we need is to determine λ for the specimen, and we get this from a parameterization based on many experimental measurements (see Malis *et al.* 1988). The expression is

$$\lambda = \frac{106\, F E_0}{\left\{ \mathcal{E}_m \ln\left(\frac{2\beta E_0}{\mathcal{E}_m}\right) \right\}} \qquad [39.13]$$

where λ is in nm, E_0 in keV, β in mrad, F is a relativistic correction factor, and $\mathcal{E}_m$ is the average energy loss in eV which, for a material of average atomic number Z, is given by

$$\mathcal{E}_m = 7.6\, Z^{0.36} \qquad [39.14]$$

The relativistic factor (F) is given by

$$F = \frac{\left\{ 1 + \dfrac{E_0}{1022} \right\}}{\left\{ 1 + \left(\dfrac{E_0}{511}\right)^2 \right\}} \qquad [39.15]$$

You can easily store these equations in the TEM computer or in your calculator and they give an accuracy for t of better than ±20%.

If indeed your specimen is so thin that only single scattering occurs, then you can use a similar expression but assume that the only significant scatter was a single plasmon event. Thus

$$t = \lambda_p \frac{I_p}{I_0} \qquad [39.16]$$

where λ_p is the plasmon mean-free path (see Table 38.2), I_p is the intensity in the first (and only) plasmon peak, and I_0 is the intensity in the zero-loss peak.

The method has advantages over other thickness measurement techniques in that you can apply it to any specimen, amorphous or crystalline, over a wide range of thicknesses.

If plural scattering is significant, then your quantification results become unreliable.

> A typical ball-park figure is that, if the intensity in the first plasmon peak is greater than one-tenth the zero-loss intensity, then your specimen is too thick.

Another way of saying this is that, if $t > 0.1\lambda_p$, then errors $> \sim 10\%$ are expected, as shown in Figure 39.12. Of course,

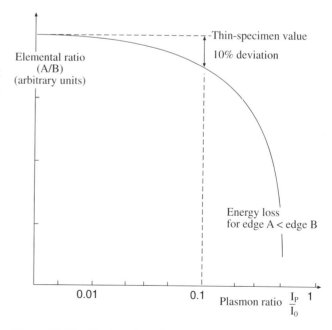

Figure 39.12. The intensity ratio of two ionization edges (*A/B*) as a function of specimen thickness. The thickness is plotted in terms of the ratio of the plasmon to the zero-loss intensity (I_p/I_0). The intensity ratio is affected significantly when I_p is above about $0.1\, I_0$.

one way round this problem is to use very thin foils, but often you can't produce thin enough specimens. Murphy's law says that the area you're interested in will usually be too thick. Then you have to deconvolute the spectra to make the single-scattering assumption valid.

39.6. DECONVOLUTION

We saw back in Figure 38.5 that the effect of plural scattering is to add intensity to the ionization edge, mainly as a result of combined inner- and outer-shell losses.

> We can represent the experimental ionization edge as a true single-scattering (hydrogenic) edge convoluted with the plasmon, or low-loss, spectrum.

The aim of the deconvolution process therefore, as shown schematically in Figure 39.13, is to extract the single-scattering intensity distribution. We'll describe two methods, the Fourier-Log and the Fourier-Ratio, which are both based on the work of Egerton *et al.* (1985), and both methods are incorporated in the Gatan ELP proprietary software. Strictly speaking, the deconvolution should be carried out in both the energy dimension and the angular dimension, but in practice all the routines ignore the angular dimension; this simplification introduces a small systematic error into any deconvolution. The error is usually <10% up to typical energy losses below about 1 keV, so we can usually ignore it. A smaller β increases the deconvolution error, since the plural-scattered electrons have a wider angular distribution and so more of them are excluded as β decreases.

The *Fourier-Log* method removes the effects of plural scattering from the whole spectrum. The technique describes the spectrum in terms of the sum of individual scattering components, i.e., the zero-loss (elastic contribution) plus the single-scattering spectrum plus the double-scattering spectrum, etc. Each term is convoluted with the "instrument response function," which is a measure of how much the spectrometer degrades the generated spectrum; in the case of a PEELS, this is the point-spread junction we described in Section 37.3. The Fourier transform of the whole spectrum (F) is then given by

$$F = F(0) \exp\left(\frac{F(1)}{I_0}\right) \qquad [39.17]$$

where $F(0)$ is the transform of the elastic contribution, $F(1)$ is the single-scattering transform, and I_0 is the zero-loss intensity. So to get the single-scattering transform you take logarithms of both sides, hence the name of the technique.

Extracting the single-scattering spectrum would ideally involve an inverse transformation of $F(1)$, but this results in too much noise in the spectrum. There are various ways around this problem, the simplest of which is to approximate the zero-loss peak to a delta function. After deconvolution, you can subtract the background in the usual way, prior to quantification.

The danger of this approach is that you may introduce artifacts into the single-scattering spectrum. In particular, any gain change in a SEELS spectrum must be removed or not incorporated in the original spectrum at all. Despite the assumptions and approximations, the net result of deconvolution is often an increase in the ionization edge jump ratio. This improvement is important when you are attempting to detect small ionization edges from trace elements, or the presence of edges in spectra from thick specimens. An example of Fourier-Log deconvolution is shown in Figure 39.14.

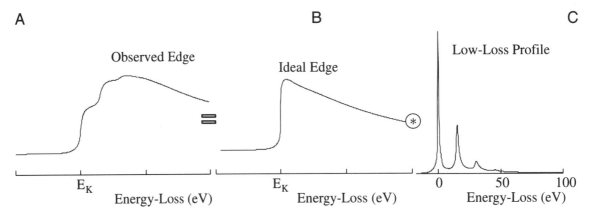

Figure 39.13.   The contribution of plural scattering to the experimentally observed ionization edge intensity profile (A) is determined by the convolution of the ideal single-scattering ionization edge (B) with the low-loss plasmon region (C).

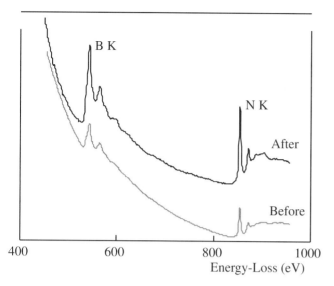

Figure 39.14. A spectrum from a thick crystal of BN before and after Fourier-Log deconvolution. The jump ratio is increased in the deconvoluted spectrum.

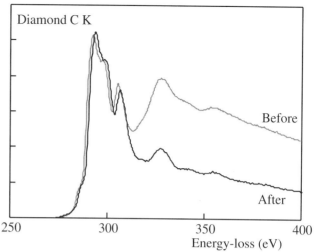

Figure 39.15. A carbon K edge from a thick specimen before and after Fourier ratio deconvolution. The plural scattering plasmon contribution to the post-edge structure is removed.

The *Fourier-Ratio* technique approximates the experimental spectrum to the ideal single-scattering spectrum, convoluted with the low-loss spectrum. We define the low-loss portion of the spectrum as the region up to ~50 eV, including the zero-loss peak, but before the appearance of any ionization edges. So we can now write

$$F' = F(1).F(P) \qquad [39.18]$$

where F' is the Fourier transform of the experimental intensity distribution around the ionization edge and $F(P)$ is the Fourier transform of the low-loss (mainly plasmon) spectrum. In this equation, therefore, the instrument response is approximated by the low-loss spectrum rather than the zero-loss peak. If we rearrange equation 39.18 to give a ratio (hence the name of the technique)

$$F(1) = \frac{F'}{F(P)} \qquad [39.19]$$

we now obtain the single-scattering distribution by carrying out an inverse transformation. In contrast to the Fourier-Log technique, you must subtract the background intensity before you deconvolute. Again, to avoid the problem of increased noise, it is necessary to multiply equation 39.19 by the transform of the zero-loss peak. Figure 39.15 shows a carbon K edge before and after Fourier-Ratio deconvolution.

Multiple Least-Squares Fitting: If your specimen is not uniformly thin, Fourier techniques won't work. Then you should use multiple least-squares (MLS) fitting of convoluted reference spectra (Leapman 1992). A single-

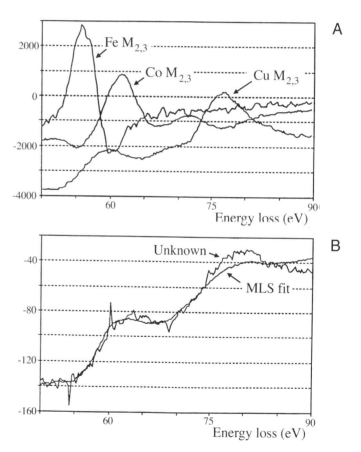

Figure 39.16. (A) Three first-difference M edge reference spectra from Fe, Co, and Cu. (B) MLS fit of the reference spectra superimposed on a low-energy portion of an experimental spectrum from an intermetallic particle in a Cu alloy showing the good fit that can be obtained.

scattering reference spectrum $R_0(\mathcal{E})$ in the region of the edge to be quantified is convoluted with the first plasmon-loss portion of the unknown spectrum (P) and the resultant spectrum $R_1(\mathcal{E}) = P*R_0(\mathcal{E})$ is used to generate several reference spectra ($R_2(\mathcal{E}) = P*R_1(\mathcal{E})$, etc.). These reference spectra are then fitted to the experimental spectrum using MLS routines and specific fitting parameters are obtained. An experimental set of Fe, Co, and Cu reference spectra is shown in Figure 39.16A and the actual fit to part of the experimental spectrum from an intermetallic in a Cu-Be-Co alloy is shown in Figure 39.16B.

In summary, to quantify ionization-loss spectra you need a single-scattering spectrum, which can be approximated if you have very thin specimens or generated by deconvolution of your experimental spectrum. It is arguable that all spectra should be deconvoluted prior to quantification, but the uncertain effects of the possible errors introduced by deconvolution mean that you should do this cautiously. Often you'll find it useful to deconvolute the point-spread junction from all PEELS spectra, since this sharpens the edge onset and any ELNES intensity variations.

Always check the validity of the deconvolution routine by applying it to spectra from a known specimen obtained over a range of thickness.

39.7. CORRECTION FOR CONVERGENCE OF THE INCIDENT BEAM

If you're working in STEM mode to get high spatial resolution, then it is possible that the beam-convergence angle, 2α, may introduce an error into your quantification. When 2α is equal to or greater than 2β, convergence effects can limit the accuracy because the experimental angular distribution of scattered electrons will be wider than expected. Therefore, you have to convolute the angular distribution of the ionization-loss electrons with the beam-convergence angle. Joy (1986b) proposed handling this through a simple equation which calculates the effective reduction (R) in $\sigma(\beta\Delta)$ when α is greater than β

$$R = \frac{\left[\ln\left(1 + \dfrac{\alpha^2}{\theta_E{}^2}\right)\beta^2 \right]}{\left[\ln\left(1 + \dfrac{\beta^2}{\theta_E{}^2}\right)\alpha^2 \right]} \qquad [39.20]$$

where θ_E is the characteristic scattering angle. So you can see that if α is small (particularly if it is smaller than β), then R is <<1 and the effect of beam convergence is negligible.

39.8. THE EFFECT OF THE SPECIMEN ORIENTATION

In crystalline specimens, diffraction may influence the intensity of the ionization edge. This effect may be particularly large if your specimen is oriented close to strong two-beam conditions. Both X-ray emission and ionization-loss intensity can change because of electron channeling effects close to the Bragg condition. At the Bragg condition the degree of beam–specimen interaction increases, compared with zone-axis illumination where no strong scatter occurs; the energy-loss processes behave similarly. This phenomenon, known as the Borrmann effect in XEDS (see Section 35.8) is not important for low-energy edges, but intensity changes of a factor of two have been reported for Al and Mg K edges (Taftø and Krivanek 1982). The use of large α minimizes the problem in XEDS, but beam-convergence effects are themselves a problem in EELS. The easiest way to avoid orientation effects is simply to operate under kinematical conditions and stay well away from any bend centers or bend contours, just as in XEDS.

39.9. SPATIAL RESOLUTION

In contrast to the situation in XEDS, beam spreading is not a major factor in determining the source of the EELS signal and so the many factors that influence beam spreading are mainly irrelevant. The spectrometer only collects those

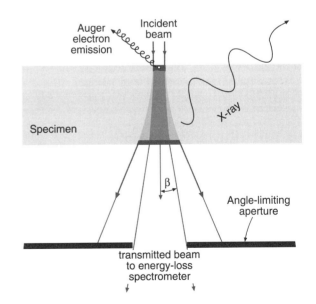

Figure 39.17. The effect of the spectrometer collection angle is to limit the contribution to the spectrum from high-angle scattered electrons, thus ensuring a high spatial resolution signal.

electrons emanating from the specimen in a narrow cone, as shown in Figure 39.17. Therefore, energy-loss electrons that are elastically scattered through large angles are excluded from contributing to your spectrum. Remember that for XEDS these same high-angle electrons would still generate X-rays some distance from the incident probe position, and these X-rays would be detected by XEDS.

In the absence of a contribution from beam spreading, the spatial resolution of ionization-loss spectrometry depends on the mode of analysis:

- The factor controlling the resolution in STEM mode, or in a probe-forming mode on a TEM, is mainly the size of the probe; we can easily get data with probe sizes <10 nm, and <1 nm with more difficulty.
- When we operate in TEM mode, the spatial resolution is a function of the selecting aperture, i.e., the spectrometer entrance aperture and its effective size at the plane of the specimen.

In TEM mode, lens aberrations usually limit the spatial resolution, as we showed back in Section 37.4.C. We know that the EELS signal isn't affected much by beam spreading and we can easily limit the source of the signal to a few nanometers with an FEG. So, there have been correspondingly fewer studies of the limits of spatial resolution. Most work on defining the spatial resolution has been pursued in France by Colliex and co-workers, e.g., Colliex (1985). Because the primary factor when operating in STEM mode (especially with an FEG) is the incident-probe diameter, we have to be concerned about the problems of spherical aberration broadening the probe (Colliex and Mory 1984). So you must be careful in your selection of the beam-defining aperture. We discussed this topic in detail in the section on the spatial resolution of XEDS.

One factor that we often consider in EELS, but ignore in XEDS (although it occurs in X-ray generation also), is the phenomenon of delocalization.

> Delocalization is the ejection of an inner-shell electron by the passage of a high-energy electron some distance from the atom.

If you are physics oriented and want to read more about this, see Muller and Silcox (1995). The scale of this wave-mechanical effect is small, in the range 2–5 nm, and it is inversely proportional to the energy loss. So it appears that, except in rare cases, delocalization will not limit the spatial resolution; the practical factors such as probe aberrations, signal-to-background in the EELS signal, and damage are much more important. So we can conclude that spatial res-

olution for EELS will be somewhat better than for XEDS and experiments seem to indicate that this indeed is the case (Colliex 1985). In fact, in certain zone-axis misorientations, it appears that the (FEG) electron beam, if it is <1 nm, can be localized to individual rows of atoms, producing atomic-level spatial resolution (see Figure 40.5C and Browning *et al.* 1993).

39.10. DETECTABILITY LIMITS

The detectability limits for ionization-loss spectrometry are governed by the same factors as we discussed for XEDS. Therefore, the inverse relationship with spatial resolution also applies. Clearly we have to optimize several factors:

- The edge intensity.
- The signal-to-background ratio (jump ratio).
- The efficiency of signal detection.
- The time of microanalysis.

EELS has an inherently higher efficiency than XEDS, but a correspondingly poorer signal-to-background because of the higher background in the spectrum. Joy (1986a) has attempted to compare the two techniques in some detail and calculations based on a thermionic source. He concluded that the MMF for EELS would be of the order of 1–10% in a Si foil 50 nm thick; this value is somewhat worse than the experimental data for XEDS in similar specimens. Leapman and Hunt (1991) have argued that in most situations, PEELS is more sensitive to the presence of small amounts of material than XEDS.

The time of collection, which strongly influences the detectability limit, is particularly dependent on whether serial or parallel collection is used. Colliex (1985) reckons

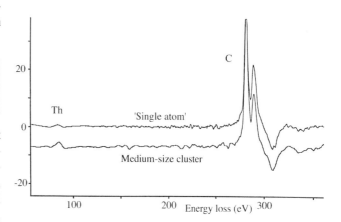

Figure 39.18. First-difference spectra showing the detection of a small cluster and a single atom of Th on a carbon support film.

that a tenfold improvement in all EELS performance criteria is to be expected if parallel collection is used. The best results, combining sensitivity and spatial resolution, will be obtained with an FEG. Krivanek *et al.* (1991) used an FEG DSTEM, parallel detection, and sophisticated data processing to detect the presence of single atoms of Th on thin carbon films, as shown in Figure 39.18. While this is a most favorable analysis situation because of high *Z* of the atoms and the low average *Z* of the support film, the result

still shows clearly the superiority of the best possible EELS microanalysis over XEDS, which cannot yet detect single atoms.

In conclusion, microanalysis using ionization edges, while considerably more difficult to perform than XEDS, appears to offer both improved spatial resolution and analytical sensitivity. Parallel collection is significantly better than serial collection in both aspects. As was the case for XEDS, an FEG source is required for the best performance.

CHAPTER SUMMARY

The ionization edges can be used to give quantitative elemental analyses from all the elements in the periodic table using a ratio equation. Beware, however, of the many experimental variables you have to define for your TEM, the PEELS, and the specimen. Compared to XEDS there have been very few quantitative analyses or composition profiles measured using EELS.

To use Egerton's ratio equation:

■ You have to subtract the background using a power law or MLS approach. The former is easier.
■ Integrate the edge intensity. That's straightforward.
■ Then you have to determine the partial ionization cross section $\sigma_K(\beta\Delta)$. This is more difficult.
■ Calculate $\sigma_K(\beta\Delta)$ with SIGMAK and SIGMAL for most K and L edges.
■ For M edges and for the lightest elements (e.g., Li), use known standards.

The difficulty with using standards is that the specimen thickness has to be the same as the unknown and the standard also has to have the same bonding type as the "unknown." This is often impossible, although specimen thicknesses can be deduced directly from the low-loss spectrum intensity. The biggest limitation to quantification is that, ideally, your specimens have to be less than one mean-free path in thickness (typically < 50 nm) otherwise deconvolution routines are needed, which can introduce artifacts on their own.

Spatial resolution and minimum detectability are better than XEDS. Single-atom detection has been demonstrated.

REFERENCES

General References

Egerton, R.F. (1996) *Electron Energy-Loss Spectroscopy in the Electron Microscope,* 2nd edition, Plenum Press, New York.
Joy, D.C. (1986a) in *Principles of Analytical Electron Microscopy* (Eds. D.C. Joy, A.D. Romig Jr., and J.I. Goldstein), p. 249, Plenum Press, New York.
Joy, D.C. (1986b) in *Principles of Analytical Electron Microscopy* (Eds. D.C. Joy, A.D. Romig Jr., and J.I. Goldstein), p. 277, Plenum Press, New York.
Maher, D.M. (1979) in *Introduction to Analytical Electron Microscopy* (Eds. J.J. Hren, J.I. Goldstein, and D.C. Joy), p. 259, Plenum Press, New York.
Williams, D.B. (1987) *Practical Analytical Electron Microscopy in Materials Science,* 2nd edition, Philips Electron Optics Publishing Group, Mahwah, New Jersey.

Specific References

Ahn, C.C. and Krivanek, O.L. (1983) *EELS Atlas,* Gatan, Inc., 780 Commonwealth Drive, Warrendale, Pennsylvania 15086.
Browning, N.D., Chisholm, M.F., and Pennycook, S.J. (1993) *Nature* **366,** 143.
Colliex, C. (1984) *Advances In Optical and Electron Microscopy* **9** (Eds. R. Barer and V. E. Cosslett), p. 65, Academic Press, New York.
Colliex, C. (1985) *Ultramicroscopy* **18,** 131.
Colliex, C. and Mory, C. (1984) *Quantitative Electron Microscopy* (Eds. J.N. Chapman and A.J. Craven), p. 149, SUSSP Publications, Edinburgh, Scotland.
Disko, M.M. (1986) in *Microbeam Analysis-1986* (Eds. A.D. Romig Jr. and W.F. Chambers), p. 429, San Francisco Press, San Francisco, California.
Egerton, R.F. (1979) *Ultramicroscopy* **4,** 169.

Egerton, R.F. (1981) in *Proc. 39th EMSA Meeting* (Ed. G.W. Bailey), p. 198, Claitors, Baton Rouge, Louisiana.

Egerton, R.F. (1989) *Ultramicroscopy* **28**, 215.

Egerton, R.F. (1993) *Ultramicroscopy* **50**, 13.

Egerton, R.F., Williams, B. G., and Sparrow, T.G. (1985) *Proc. Roy. Soc.* **A398**, 395.

Hofer, F., Golob, P., and Brunegger, A. (1988) *Ultramicroscopy* **25**, 181.

Krivanek, O.L., Mory, C., Tence, M., and Colliex, C. (1991) *Microsc. Microanal. Microstruct.* **2**, 257.

Leapman, R.D. (1992) in *Transmission Electron Energy Loss Spectrometry in Materials Science* (Eds. M.M. Disko, C.C. Ahn, and B. Fultz), p. 47, TMS, Warrendale, Pennsylvania.

Leapman, R.D. and Hunt, J.A. (1991) *Microsc. Microanal. Microstruct.* **2**, 231.

Liu, D.R. and Williams, D.B. (1989) Proc. Roy. Soc. London **A425**, 91.

Malis, T., Rajan, K., and Titchmarsh, J.M. (1987) in *Intermediate Voltage Electron Microscopy* (Ed. K. Rajan), p. 78, Philips Electron Optics Publishing Group, Mahwah, New Jersey

Malis, T. and Titchmarsh, J.M. (1988) in *Electron Microscopy and Analysis–1985* (Ed. G.J. Tatlock), p. 181, Adam Hilger Ltd., Bristol and Boston, Massachusetts.

Malis, T., Cheng, S., and Egerton, R.F. (1988) *J. Electron Microsc. Tech.* **8**, 193.

Muller, D.A. and Silcox, J. (1995) *Ultramicroscopy* **59**, 195.

Rez, P. (1989) *Ultramicroscopy* **28**, 16.

Taftø, J. and Krivanek, O.L. (1982) *Phys. Rev. Lett.* **48**, 560.

Zaluzec, N.J. (1981) in *Analytical Electron Microscopy-1981* (Ed. R.H. Geiss), p. 193, San Francisco Press, San Francisco, California.

Everything Else in the Spectrum

40

CHAPTER PREVIEW

The energy resolution of the magnetic prism spectrometer is very good, which means that the energy-loss spectrum contains a wealth of information about the specimen in addition to its basic elemental chemistry. In the previous chapter, we mentioned how we can learn about chemistry using ionization edges. Much of this chemical information is contained in fine-detail intensity variations at the ionization edges in the core-loss spectra termed *energy-loss near-edge structure* (ELNES) and *extended energy-loss fine structure* (EXELFS). From this fine structure, we can obtain information on how the ionized atom is bonded, the coordination of the atom, and its density of states. Furthermore, we can probe the distribution of other atoms around the ionized atom, i.e., the radial distribution function (RDF). Understanding these phenomena requires that we use certain concepts from atomic and quantum physics. The nonphysicist can skip some sections at this time and just concentrate on the results. The rewards of working through this topic will be an appreciation of some of the more powerful aspects of EELS.

> If high spatial resolution is important, you can't obtain this additional information by any other spectroscopic technique.

In addition to the extra information around the ionization edges, we can extract useful data from the low-loss region (<50 eV) of the spectrum. The predominant features in this part of the spectrum are the plasmon peaks, which represent the response of the weakly bound valence and conduction electrons to the high-energy incident electron. The plasmon response contains direct information about the free-electron density. In some binary free-electron alloys, plasmon-peak shifts reflect the composition of the specimen. Within the low-loss region, but separate from the intense plasmon peak, we can find intensity that is related to the dielectric constant of the specimen. Furthermore, we can discern certain inter/intraband transitions, especially in polymers, and we can measure directly the band gap of semiconductors and insulators. We also introduce briefly the effect of the angle of scatter of the energy-loss electrons, which can be studied using the DP.

We note how the intense nature of the EELS spectrum, due mainly to the very high collection efficiency of the spectrometer, permits EELS imaging. Energy-loss (or energy-filtered) images and DPs can be formed in two ways: slowly, quantitatively, and digitally, or rapidly and qualitatively in an analog fashion. The primary advantage of EELS imaging is that *all* the information available in the spectrum can be imaged and related to all the other diffraction and imaging techniques that come from the TEM.

Everything Else in the Spectrum

40

40.1. FINE STRUCTURE IN THE IONIZATION EDGES

We saw in Section 38.4 that the ionization edges have intensity variations both within about 30 eV of the onset of the edge (ELNES) and extending for several hundred eV as the edge intensity diminishes (EXELFS). This fine structure contains a wealth of useful information, but to understand its origins you have to use some ideas from quantum physics.

> Both ELNES and EXELFS arise because the ionization process can impart more than the critical ionization energy (E_c) needed by the core electron to escape the attraction of the nucleus.

Any excess energy ($>E_c$) that the core electron possesses can be imagined as a wave emanating from the ionized atom. So again, we have to switch from a particle to a wave model of the electron, as we've done before, e.g., when we talked about diffraction in Part II. If this wave has only a few eV of excess energy, it undergoes plural elastic scattering from the surrounding atoms, as shown schematically in Figure 40.1A; this scattering is responsible for the ELNES, as we'll show. If the wave has even more excess energy, then it is less likely to be scattered several times and we can approximate the cause of the EXELFS to a single-scattering event, as shown in Figure 40.1B. Thus, EXELFS and ELNES can be viewed as a continuum of electron-scattering phenomena, with the arbitrary distinction that ELNES is confined to a few tens of eV past the edge onset. While ELNES arises from a more complex process than EXELFS, it is more widely used, because the ELNES is more intense, and so we'll discuss it first.

40.1.A. ELNES

The Physics: A core electron may receive enough energy from the beam electron to be ejected, but not enough to es-

cape to the vacuum level. So it is still not free of all specific nuclear attraction. In such circumstances, the final state of the core electron will be in one of a range of possible energy levels above the Fermi energy (E_F). You may recall that the Fermi level, or the Fermi surface in three dimensions, is the boundary between the filled states and the unfilled states in the weakly bound conduction/valence bands (although, strictly speaking, this statement is only true when $T = 0$ K). In a metal, there is no separate valence band and E_F sits somewhere in the conduction band, as shown schematically in the classical energy level diagram of an atom in Figure 40.2. In an insulator or a semiconductor, E_F is between the valence band (which has all filled states) and the conduction band (which has no filled states).

The EELS: The excited electron can reside in any of the unfilled states, but not with equal probability. Some empty states are more likely to be filled than others because there are more states within certain energy ranges than in others. This uneven distribution of electron energy levels is termed the density of states (DOS) and this is also shown in the right diagram in Figure 40.2. Because of the greater probability of electrons filling certain unoccupied states above E_F, the intensity in the ionization edge is greater at the corresponding energy losses above the critical ionization energy E_C (which is equivalent to E_F), as shown in Figure 40.3.

> This variation in intensity, extending several tens of eV above E_C, is the ELNES and is effectively a probe of the DOS above E_F.

The Application: The importance of ELNES is that the DOS is extremely sensitive to changes in the bonding, or the valence state, of the atom. For example, if you look ahead to Figure 40.5 the carbon K ELNES is different for graphite and diamond and the Cu L ELNES changes when Cu is oxidized to CuO. On an even more detailed level, we can deduce the coordination of the ionized atom from the shape of the ELNES.

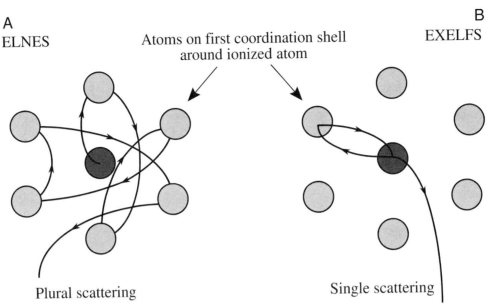

Figure 40.1. Schematic diagram showing the source of (A) ELNES and (B) EXELFS. The excess energy above the ionization threshold creates a wave radiating from the ionized atom which is scattered by surrounding atoms. The low-energy ELNES arises from multiple scatter and is affected by the bonding between the atoms. The higher-energy EXELFS approximates to single scatter and is affected by the local atomic arrangement.

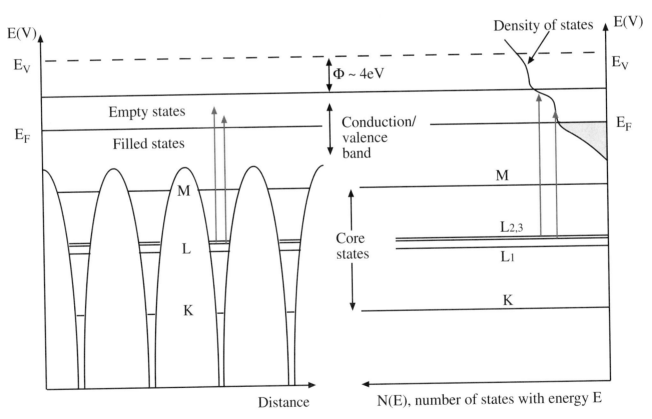

Figure 40.2. Relationship between the classical energy diagram of a metal atom (left) and the density of filled (shaded) and empty (unshaded) states (DOS) in the conduction band (right). The DOS is approximately a quadratic function on which small variations are superimposed. Ionization results in electrons ejected from the core states into empty states above the Fermi level (E_F).

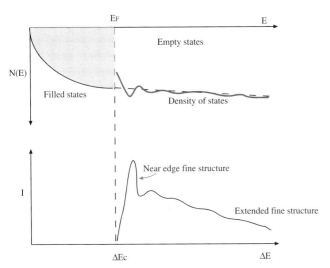

Figure 40.3. Relationship between the empty DOS and the ELNES intensity. Note the equivalence between the Fermi energy E_F and the ionization edge onset E_C. Electrons ejected from the inner shells reside preferentially in regions of the DOS with the greatest density of empty states. The filled states below E_F are shown as a quadratic function, but this is an approximation.

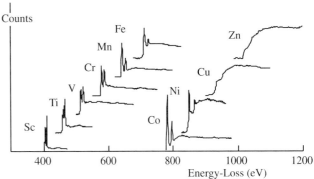

Figure 40.4. The L_3 and L_2 white lines in spectra from the transition metals show a slow variation in intensities until Cu, which has no white lines because the d shell is full.

> Even if you don't understand the intricacies of the DOS and Fermi surfaces, you can still deduce bonding information simply by comparing your experimental ELNES with that from standard specimens of known valence state or coordination.

The EELS Atlas, which we've already referred to several times, contains many oxide spectra as well as elemental ones.

Perhaps the most startling example of ELNES is the presence of the "white lines," which we introduced in Section 38.4. They are intense sharp peaks on certain ionization edges; the L edges of the transition metals show such lines. Reminder: white lines were first seen on photographic plates in early X-ray absorption spectroscopy experiments. The white lines in the transition metal L edges are the L_3 and L_2 edges, respectively, as shown in Figure 40.4A. We'll explain what happened to L_1 later. To explain these lines we need a little more quantum physics, which you can skip if you wish and go to the last paragraph of this section. You should also be aware that there is disagreement as to whether white lines are truly "fine structure" or strictly ionization edge (atomic) intensity; but we'll not discuss this somewhat arcane argument.

More Physics: First, go back and look at Figure 38.4 to remind yourself that the various electron energy levels, K, L, M, etc., correspond to principal quantum numbers (n) equal to 1, 2, 3, etc. Within those energy levels, the electrons may have s, p, d, or f states, for which the angular momentum

quantum number (ℓ) equals 0, 1, 2, 3, respectively. The notation s, p, d, f comes from the original description of the atomic spectral lines arising from these electron states, namely, sharp, principal, diffuse, and fine, although these have no counterpart in the EELS spectra we obtain.

> As we noted in Section 38.4, the nomenclature $L_{2,3}$ arises from the fact that the L shell, from which the electron was ejected, has different energy levels. Such separation of the energies of the core states is called spin–orbit splitting.

Because the L electrons in levels 2 and 3 are in the p state, quantum theory demands that the sum (j) of their spin quantum number (s) and angular momentum quantum numbers (ℓ) is governed by the Pauli exclusion principle such that j ($= s+\ell$) can only equal 1/2, 3/2, 5/2, etc. The spin quantum number, s (not to be confused with the s state), can only equal ±1/2. Taking all this into account along with other quantum number restrictions, it turns out that in the higher-energy (more tightly bound) L_2 shell we can have 2 p electrons with $j = \pm1/2$ while in the L_3 shell we can have 4 p electrons with $j = \pm1/2, \pm3/2$. Therefore, we might expect twice as many electrons to be excited from the L_3 shell as from the L_2 shell, giving an L_3/L_2 intensity ratio of 2. While this rule is approximately obeyed in the Fe spectrum only, in practice the ratio is seen to increase along the transition metal series from 0.8 for Ti to 3 for Ni, as is also seen in the spectral sequence in Figure 40.4.

Now these p-state electrons in the L shell cannot be excited to just any unoccupied state.

> The change $\Delta\ell$ in the angular momentum quantum number between the initial and final states must equal ±1. This constraint is called the *dipole selection rule.*

So for the p state ($\ell = 1$) the only permitted final states are either an s state ($\ell = 0$) or a d state ($\ell = 2$). Consequently, the electrons go up primarily into the unoccupied d states, since there are very few unfilled s states in the conduction band.

It is because of the dipole selection rules that we don't see a strong L_1 edge in the spectrum. The L_1 edge sits closer to the nucleus than the L_2 and L_3 edges and its electrons are in the s state ($\ell = 0$) so they can only be excited to a p state ($\ell = 1$), but not to a d state ($\ell = 2$), or to another s state. Since there are few unfilled p states in the conduction band of transition metals and they are much more spread out in energy than the d states, the L_1 intensity is very low and the peak is broad and may be invisible in the $L_{2,3}$ post-edge structure.

In fact, the energy width of the white lines is also affected by the time it takes for the ionized state to decay. One form of Heisenberg's uncertainty principle states that $\Delta E \Delta t = h/4\pi$, so a rapid decay gives a wide peak. For example, the Fe L_2 ionization can be rapidly compensated by an electron from the L_3 shell filling the hole and ejecting an Auger electron from the d shell. (This is called a Coster–Kronig transition.) A conduction-band electron could also fill the L_2 core hole, but the L_3 core hole can *only* be filled from the conduction band. Therefore, because there are two possible ways to fill the L_2 core hole, the L_2 line has a shorter Δt and a larger ΔE than the L_3 line, which is much sharper.

Back to Applications: So let's see how all of this can be useful. If you look at Figure 40.5A you'll see the carbon K edges for graphite and diamond. The carbon atom has hybridized s and p orbitals (termed σ and π in molecular-orbital theory). Graphite contains sp^2 bonds in the basal plane with van der Waals bonding between the planes. The diamond structure, in contrast, has four directional hybridized sp^3 covalent bonds. In diamond, atoms are tetrahedrally coordinated rather than arranged in graphite sheets. The strong peak on the rising portion of the K edge identifies the empty π^* states into which the K-shell electrons are transferred in graphite, while the diamond K edge has no such peak. This kind of information is extremely useful in the study of thin diamond and diamond-like carbon films, which are of tremendous current interest to both semiconductor manufacturers and the coatings industry. Carbon films can be made with a continuous range of graphitic and diamond-like character and it is possible to deduce the relative fraction of sp^3 (diamond) and sp^2 (graphite) bonding from the K-edge ELNES (Bruley *et al.* 1995). Another useful example is given in Figure 40.5B, where the changes in the Cu $L_{2,3}$ edge with oxidation are shown. This is a classical example, since Cu metal has all its 3d states filled so there are no white lines in spectra from the metal. Upon oxidation, some 3d electrons are transferred to the oxygen, leaving unfilled states, and the white lines appear in the oxide spectrum. Note also that the onset of the oxide edge is

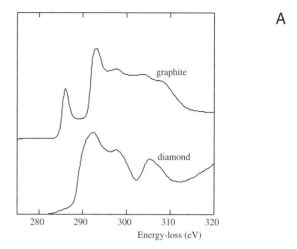

A

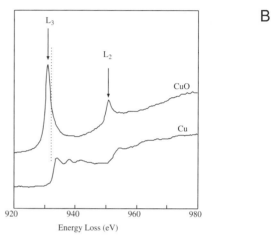

B

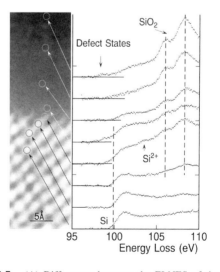

C

Figure 40.5. (A) Differences between the ELNES of the carbon K edge from graphite and diamond, (B) change in the Cu L edge as Cu metal is oxidized, (C) change in ELNES of the Si $L_{2,3}$ K edge in a series of spectra gathered from individual atom rows across the interface between crystalline Si and amorphous SiO_2.

different from that of the metal, because this electron transfer changes the value of E_C.

> This phenomenon is called a chemical shift and also helps to fingerprint the specimen.

Finally, in Figure 40.5C the Si L edge ELNES is seen to change across a Si-SiO$_2$ interface because the Si bonding changes. In this example, you can see the extraordinary power of an FEG STEM to provide simultaneous atomic-level images and spectra localized to individual atomic columns.

> The combination of Z-contrast imaging (see Section 22.4) and PEELS is arguably the most powerful analytical technique for atomic characterization (see, e.g., Batson 1995, Browning and Pennycook 1995).

ELNES Calculations: Many attempts have been made to compare the ELNES with calculations of the density of states in simple materials such as metals and oxides. While the experimental and calculated spectra show reasonable agreement in terms of the energy of the various spectral features, there are still some discrepancies in the measured and calculated intensities. Great strides have been made in the last few years, mainly in improvements in models of the atomic potentials and in the computing power needed to pursue the calculations. This aspect is transforming the study of ELNES from an esoteric field to one with broad applications in materials science. There are two approaches to calculating the ELNES:

■ Calculate the band structure directly in reciprocal space.
■ Calculate the effect of multiple scattering of the electron wave in real space using the model shown in Figure 40.1A.

It can be shown that in fact these two approaches are mathematically equivalent. We'll emphasize the latter method since it is more commonly used.

Various approximations are made in ELNES modeling. The most critical approximation arises from the choice of the atomic potential. A common choice is the so-called "muffin-tin potential" in which a constant potential is assumed in the regions between atoms that don't touch. The potential within the atom must be spherically symmetrical. This model modifies the classical energy diagram, as shown in Figure 40.6. Apparently, this energy profile approximates to a cross section of a tin used by physicists to bake muffins. The potential profile across dissimilar adjacent atoms is asymmetrical, as also shown in Figure 40.6.

Having chosen a potential, the ELNES is determined by calculating all possible inter- and intra-shell scattering events suffered by the electron after it emerges above the Fermi level. A similar calculation is made in X-ray absorption near-edge structure (XANES) studies, and the two phenomena are equivalent. The wavelength is governed by the excess electron energy ($E-E_F$). If the wavelength is long (i.e., low energy), the electron is scattered many times by the first few shells of atoms surrounding the ionized atom. One of the problems that confuses the issue is that the ion-

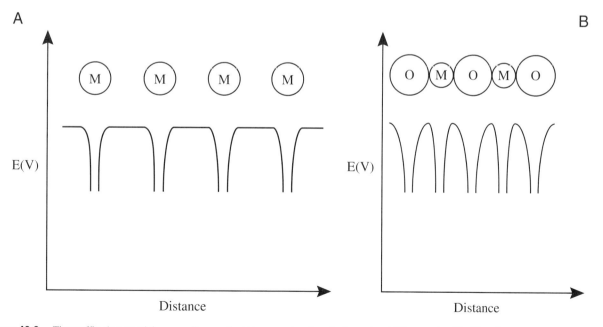

Figure 40.6. The muffin-tin potential energy diagram for (A) a non-closed-packed metal and (B) a metal oxide. Note the symmetry of the potential wells for the metal and the asymmetry for the oxide.

ization event results in a hole in the core shell, thus changing the atomic potential. This is called the *core-hole effect*, and it is accounted for by approximating the ionized atom to one with a nuclear charge of Z+1, since the missing electron lowers the shielding effect of the core electrons.

> In ceramics and semiconductors the ionized electron remains localized to the ionized atom and may interact with the hole creating an electron-core hole bound state, termed an exciton. Creation of an exciton may influence the ELNES, although this remains a matter of some debate.

It can be shown that the multiple-scattering calculations predict modulations to the intensity of the ionization edge that correspond directly to the DOS of the ionized atom. So you should be aware that these calculations are only an *interpretation* of what actually happens to the electron after it emerges above the Fermi level.

Figure 40.7 shows a comparison of the calculated and theoretical ELNES for the Al L edge in tetrahedral and octahedral coordination in spinels. The difference due to different coordination is obvious. The sharp peak at the Al L-edge onset is thought to be an exciton. This effect is not well modeled by the theory, which otherwise makes a good match with the experimental data. The seminal paper in the field of ELNES experiments on transition metals and oxides is by Leapman *et al.* (1982), and a concise summary was given by Brydson (1991).

40.1.B. EXELFS

If the ejected electron does not fill an empty state, then its excess energy can also be interpreted as an electron wave

which can be diffracted by the surrounding atoms in the structure, giving rise to EXELFS. Because the electron has higher energy than those which gave rise to ELNES, the diffraction is assumed to be single scattering, as shown in Figure 40.1B. As with any diffraction event, there is information about atomic positions in the EXELFS.

> So ELNES is multiple scattering and EXELFS is single scattering, although the two phenomena overlap since the L₁ ELNES peak is often far enough past the edge onset to be included in the EXELFS.

The EXELFS modulations are each 20–50 eV wide (just visible in Figure 40.8A), and continue for several hundred eV. EXELFS is exactly analogous to the oscillations seen in the extended X-ray absorption edge fine structure (EXAFS) in synchrotron X-ray spectra. However, EXAFS results from complete photoabsorption of the incident X-ray while EXELFS involves absorption of only a small fraction of the energy of the beam electron.

Experimentally, it's not easy to see the EXELFS modulations because they are only about 5% of the edge intensity, and so you need good counting statistics. With SEELS you may have to gather the spectrum for many minutes or even hours, so PEELS is the only realistic way to pursue EXELFS. A thermionic source is probably best because it can deliver more current than an FEG; for this application, energy resolution is often less important. TEM diffraction mode will also increase your total signal intensity. Either way, you pay a price in terms of a loss of spatial resolution and an increased chance of specimen damage. If you need the best spatial resolution, an FEG and STEM mode is best.

We're interested in EXELFS because of the structural information contained in the intensity oscillations. To extract this information, you can use the Gatan ELP software (see Section 1.5), but you first have to ensure that the spectrum contains single-scattering information only, otherwise the plural-scattering intensity may mask the small EXELFS peaks.

> Deconvolution is always the first step if the specimen isn't thin enough, i.e., if the plasmon peak is greater than 10% of the zero-loss peak.

Next, you have to remove the background if it wasn't done prior to deconvolution. Then the EXELFS intensity modulations are fitted to a smooth curve, and the intensity either side of the curve is plotted in **k** space (reciprocal space) (Figure 40.8B)

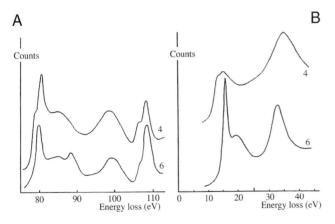

A   B

Figure 40.7. Comparison of (A) experimental spectra and (B) theoretical ELNES calculations for the Al L$_{2,3}$ edge in tetrahedrally coordinated (CN-4) and octahedrally coordinated minerals (CN-6). The calculated energy axis in (B) refers to eV above the L edge onset of ~75 eV.

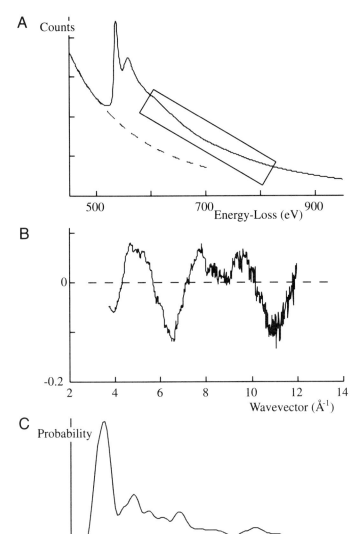

Figure 40.8. (A) EXELFS modulations are barely detectable in the selected post-edge region of an ionization edge. (B) The oscillations either side of a curve tilted to the post-edge spectrum are plotted in k space before (C) Fourier transforming the data to produce a radial distribution function.

$$k = \frac{2\pi}{\lambda} = \frac{\left[2m_0(E - E_K)\right]^{\frac{1}{2}}}{h} \qquad [40.1]$$

where E_K is the edge onset energy, E is the energy of the ejected electron, of wavelength λ, and the rest of the terms have their usual meaning. The electron wave interference gives periodic intensity maxima in **k** space when

$$\left(\frac{2a}{\lambda}\right)2\pi + \Phi = 2\pi n \qquad [40.2]$$

Here a is the distance from the ionized atom to the first scattering atom, and Φ is the phase shift that accompanies the scattering. Therefore, there are periodic maxima occurring for $n = 1$, 2, etc., and for different interatomic spacings. Consequently, you should be able to determine the local atomic environment, if the various interferences can be discriminated. The atomic spacing is obtained by a Fourier transform of the **k**-space modulations to give a radial distribution function, originating at the ionized atom (Figure 40.8C). Peaks in the RDF indicate the probability of an atom occurring a certain distance from the origin.

With EXELFS we can determine the partial RDF around a specific atom, and we are not restricted to the heavier atoms ($Z > 18$) needed for EXAFS. So there is great potential for studying materials such as low-Z glasses, amorphous Si, and quasicrystalline structures. The high spatial resolution is obviously advantageous and all the data can be compared with diffraction patterns and images of the analyzed area. However, like all EELS techniques, we can't get good EXELFS unless the specimen is very thin. Despite these advantages, RDF work continues to be dominated by synchrotron sources because of the intensity of the signal, but EXELFS studies are increasing, e.g., Sklad *et al.* (1992), Qian *et al.* (1995).

RDF data acquired through EXELFS complement another TEM method of acquiring RDF information. This involves energy filtering of SAD patterns by scanning the pattern across the entrance aperture to the PEELS using post-specimen scan coils (Cockayne *et al.* 1991; see also Sections 18.6 and 40.3). Effectively, a full spectrum is available at each scattering angle but, in fact, only the zero-loss (ideally only the elastic) electrons are required. The plot of the zero-loss intensity as a function of scattering angle constitutes a line profile across a filtered diffraction pattern from which the RDF can be extracted; you can see a related example if you look ahead to Figure 40.15. This process does not have the spatial resolution of EXELFS, since typical SAD patterns are integrated over ~0.2–1 µm², but the signal is much stronger than EXELFS. Accuracies of ±0.001 nm in nearest-neighbor distances can be obtained, and the process is rapid enough to be performed on-line.

The techniques of ELNES and EXELFS are really quite remarkable demonstrations of quantum theory and the wave-particle duality. Consider that within the spectrum we are only gathering beam electrons that have been scattered by the specimen atoms, yet we are able to deduce information about what happened to those atoms *after* the beam–specimen interaction and where the atoms are in the structure!

An approximate particle-based analogy would be to imagine that we are catching bowling balls that have been thrown at pins, arranged in a certain pattern. (Although instructive, this exercise is best carried out as a thought experiment!) From the velocity of the balls that we catch, we are able not only to identify the weight of the pin that was hit (i.e., identify the characteristic ionization edge), but also to deduce how the pin fell down and where it rolled (the ELNES). Furthermore, we can also work out the spatial arrangement of the surrounding pins that didn't fall down (the EXELFS).

So how does the beam electron know where the core electron went after it left the core shell? The answer lies in the fact that the bowling ball (particle) analogy is totally inadequate. In fact, only certain electron transitions are allowed and the beam electron can therefore only transfer certain quantized energies to the core electron, not a continuum of possible energies. So the beam electron does know the possible final state of the core electron.

40.2. THE LOW-LOSS SPECTRUM

40.2.A. Plasmon Losses

The low-energy plasmon-loss region of the spectrum also contains chemical information, because the composition of the specimen may affect the free-electron density, n, which in turn changes the plasmon-energy peak position, since the two are related, as we described back in equation 38.6. Historically, this technique was the first aspect of EELS to produce quantitative microanalysis data, and it has been used in a limited number of systems, mainly aluminum and magnesium alloys in which the plasmon-loss spectrum is dominant and consists of sharp Gaussian peaks. For a review see Williams and Edington (1976).

The principle of plasmon-loss microanalysis is based on empirical observation of the shift in the plasmon peak position ($\mathcal{E}_P$) with composition (C), giving an expression of the form

$$\mathcal{E}_P(C) = \mathcal{E}_P(0) \pm C\left(\frac{d\mathcal{E}_P}{dC}\right) \qquad [40.3]$$

where $\mathcal{E}_P(0)$ is the plasmon energy loss for the pure component. By creating a series of binary alloys of known composition we can develop a working curve, which we can then use to calibrate measurements of $\mathcal{E}_P$ in unknown alloys. Table 40.1 summarizes the available plasmon-loss data for Al alloys, gathered in this manner.

Table 40.1. Alloys in Which the Variation of Plasmon Energy Loss $\mathcal{E}_p$ Has Been Measured as a Function of Composition

Alloy (at. %)	Range	$\mathcal{E}_p$ (eV) variation with fractional concentration C
Al-Mg	0–100	$\mathcal{E}_p = 15.3 - 5.0\,C_{Mg}$
Al-Mg	0–8	$\mathcal{E}_p = 15.3 - 4.4\,C_{Mg}$
Mg-Al	0–9	$\mathcal{E}_p = 10.61 + 5.9\,C_{Al}$
Al-Cu	0–2	Nonlinear
Al-Cu	0–2	$\mathcal{E}_p = 15.3^a - 10\,C_{Cu}$
Al-Cu	0–17.3	$\mathcal{E}_p = 15.3 + 4.0\,C_{Cu}$
Al-Zn	0–30	$\mathcal{E}_p = 15.3 - 0.2\,C_{Zn}$
Al-Ag	0–6	$\mathcal{E}_p = 15.3^a + 1.6\,C_{Ag}$
Al-Li	0–25	$\mathcal{E}_p = 15.3^a - 4.0\,C_{Li}$
Al-Ge	0–10	$\mathcal{E}_p = 15.3 + 0.1\,C_{Ge}$
Al-Zn-Mg	0–4	$\mathcal{E}_p = 15.3 - 4.7\,C_{Mg}$

[a]Normalized to 15.3 eV energy loss for pure Al.

Since plasmon-loss analysis demands the measurement of peak *shifts* rather than peak positions, you need an energy spectrum of the highest resolution and sufficient dispersion to measure the peak centroid accurately. The early plasmon-loss studies did not have access to FEGs and so the resolution of the thermionic source was a limiting factor. The poor resolution was compensated for to some extent by utilizing a high-dispersion Wien Filter or a Möllenstedt electrostatic spectrometer and recording the spectra photographically. More recently, similar results have been achieved using the relatively low-dispersion magnetic-prism spectrometer and electronic recording, but with an FEG. While the shift in the position of the plasmon peak may be as small as ~0.1 eV, the position of the peak centroid can still be measured to an accuracy of ~0.05 eV by computerized peak-fitting (Hunt 1995). Figure 40.9 illustrates some early plasmon-loss concentration data and the visible peak shifts that occur.

Plasmon-loss spectrometry has high spatial resolution and is relatively insensitive to specimen thickness and surface deposits. The spatial resolution is controlled by the localization of the plasmon oscillation, which is only about 10 nm, since the plasmon disturbance is rapidly damped in the free-electron gas. Your specimen thickness only affects the number and intensity of the plasmon peaks, not their position, as we described back in Figure 38.2. In fact, you get the best results from plasmon-loss spectrometry when your specimen is about 1–2 mean free paths (λ_p) thick, so that several intense Gaussian peaks are observable. The plasmon signal is intense and is the dominant loss feature in the spectrum. There are unfortunately strong practical disadvantages, which account for the almost complete absence of plasmon-loss data since the advent of ionization-loss techniques in the mid-1970s.

A

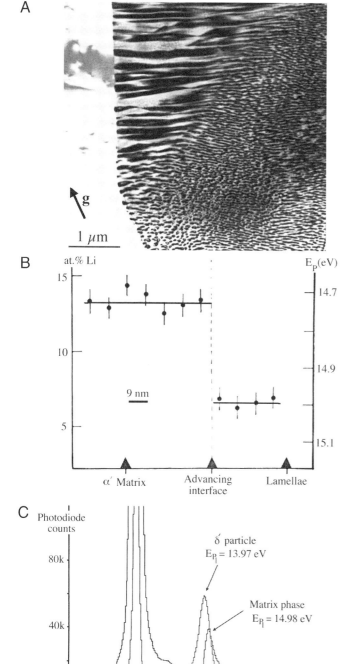

B

C

Figure 40.9. (A) Discontinuous transformation interface in Al-11 at.% Li. (B) Plasmon-loss variation and related Li composition change along the interface in (A). (C) Comparison of spectra from the matrix (5 at.% Li) and the precipitate (25 at.% Li) reveals the shift in the plasmon peak.

We are limited to specimens showing well-defined peaks, and only binary specimens can be sensibly analyzed.

In addition, the alloying element must produce a detectable change in $\mathcal{E}_p$ and this is by no means always the case. For example, the addition of 30 at.% Zn to Al scarcely changes $\mathcal{E}_p$. It is possible that application of modern detection and data-processing techniques may improve the quality and ease of acquiring and analyzing plasmon-loss spectra. However, it is not clear that they will permit the technique to be expanded significantly past the limited range of materials to which it has already been successfully applied.

While quantitative plasmon-loss microanalysis is limited, you can still use the plasmon part of the spectrum to identify unknown phases by the technique of "fingerprinting," as we showed in Figure 38.3. The low-loss portion of the spectrum is often sufficiently distinctive for different compounds that, since suitable libraries of known spectra exist that we've already referenced, you can use these libraries to cross-check the spectra from unknowns. In fact the plasmon-loss spectrum is more robust than the ionization-loss spectrum, since it will not change significantly as you change such experimental variables as α, β, kV, and it is insensitive to the data-processing variables that plague ionization-loss spectra. For example, direct examination of the low-loss spectrum is sufficient to distinguish between free-electron metals and transition metals, as shown in Figures 40.10A,B. Similarly, the low-loss spectra of the different oxides are equally distinctive (Figure 40.10C).

40.2.B. Dielectric-Constant Determination

We can view the energy-loss process as the dielectric response of the specimen to the passage of a fast electron. As a result, your energy-loss spectrum contains information about the dielectric constant or permittivity (ε). The single-scattering spectrum intensity $I(\ell)$ is related to ε by the expression (Egerton 1996)

$$I(\ell) = I_0 \frac{t}{k} \operatorname{Im}\left(-\frac{1}{\varepsilon}\right) \ln \left| 1 + \left(\frac{\beta}{\theta_E}\right)^2 \right| \qquad [40.4]$$

where I_0 is the intensity in the zero-loss peak, t is the specimen thickness, and k is a constant incorporating the electron momentum and the Bohr radius. You can use a Kramers–Kronig analysis to analyze the energy spectrum in order to extract the real part of the dielectric constant from the imaginary part (Im) in equation 40.4, and details

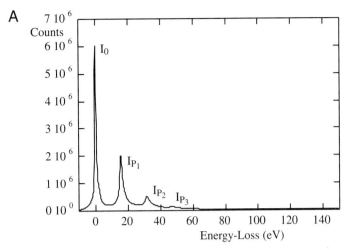

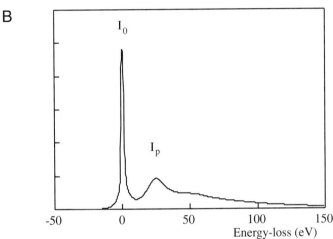

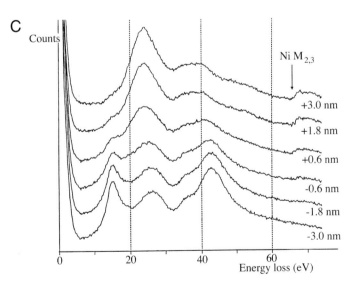

Figure 40.10. (A) Multiple plasmon peaks from Al, which is a free-electron metal, compared with (B) the single weak plasmon from a transition metal, Fe. (C) Six low-loss spectra taken across a NiO (top)–ZrO$_2$ (bottom) interface showing characteristic differences in the plasmon intensities which occurred within ±3 nm of the interface.

are given in Egerton (1996). Since you need a single-scattering spectrum, deconvolution is again the first step.

> The Kramers–Kronig analysis gives the energy dependence of the dielectric constant and other information which we usually obtain by optical spectroscopy.

The advantage of EELS for this kind of work is the improvement in spatial resolution over electromagnetic radiation techniques. Also, the frequency range which is available is more extended. The low-energy plasmon part of the energy-loss spectrum out to about 20 eV is of most interest to us, and corresponds to optical analysis of the dielectric response from the visible through the ultraviolet frequency range. So in a single EELS experiment you can, in theory, substitute for a whole battery of optical spectroscopy instrumentation. Physicists are most interested in the low-frequency range around 1 eV, since this is less accessible through optical spectroscopy. For this you need an FEG and a high-resolution spectrometer, and you need to deconvolute out the tail of the zero-loss peak so it does not mask the low-energy intensity. An example of the correspondence between EELS and optical dielectric constant spectra is shown in Figure 40.11.

40.2.C. Band-Gap and Interband Transitions

In the region of the spectrum immediately after the zero-loss peak, and before the rise in intensity preceding the plasmon peak, you can see a region of low intensity. If this intensity approaches the dark noise of the detector, then there are no electron–electron energy transfers occurring. This effect implies that there is a forbidden transition region, which is simply the band gap between the valence and conduction bands in semiconductors and insulators. Figure 40.12A illustrates the variable band gap from specimens of Si, SiO$_2$, and Si$_3$N$_4$. In this region there are sometimes small peaks that correspond to interband transition which require energy losses of <10 eV, and surface plasmons may occur. An example of an interband transition is given in the spectra from two polymers shown in Figure 40.12B.

40.2.D. Angle-Resolved EELS

Most of the time we've been talking about locating the beam at different positions on the specimen and gathering a spectrum by sending the direct beam into the spectrometer. This is often called "spatially resolved" EELS since spectra came from different spatial locations on the speci-

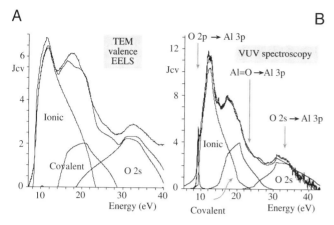

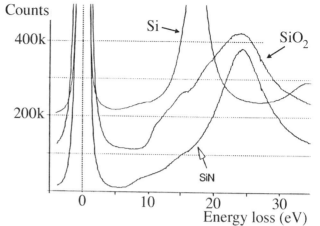

Figure 40.11. Comparison of thin-specimen EELS (A) and bulk-specimen optical dielectric constant data (B) for α-Al$_2$O$_3$. J_{cv} is the interband transition strength and the various transitions are labeled: transitions from the filled O 2p level represent ionic bonding, transitions from the hybridized Al=O level represent covalent bonding, interband transitions from O 2s–Al 3p are also detected. Individual contributions to the spectra have been obtained via a critical point model.

men. However, we have occasionally mentioned that the *angle* of scatter of energy-loss electrons is important, and there is a whole field of EELS research that studies angle-resolved spectra. To do this, we just scan the DP across the PEELS entrance aperture (or the SEELS slit) and gather spectra at different angles, as for RDF measurements that we just described. However, rather than studying the energy of electrons primarily, this technique emphasizes the determination of the *momentum* of the energy-loss electrons. Momentum transfer studies were pioneered by Silcox and co-workers (e.g., Leapman and Silcox 1979), and now with FEG STEMs you can get even more information about the symmetry of electronic states which complements spatially resolved ELNES (e.g., Wang *et al.* 1995).

One practical aspect of angle-resolved EELS is the study of Compton scattering, which is the ejection of outer-shell electrons by high-energy photons or electrons. We can detect these Compton-scattered electrons by observing the EELS spectrum at a high scattering angle (about 100 mrad), either by displacing the objective aperture to select an off-axis portion of the diffraction pattern or by tilting the incident beam. This process has been used to analyze the angular and energy distribution of Compton-scattered electrons and determine bonding information, since the Compton-scattering process is influenced by the binding energy (Schattschneider and Exner 1995).

You can appreciate now that there is a wealth of detail in the energy-loss spectrum beyond the basic chemistry of the specimen. To extract this information you need a single-scattering (deconvoluted) spectrum and sophisticated mathematical analysis. Often, our interpretation of the data is limited by lack of knowledge of the physics of

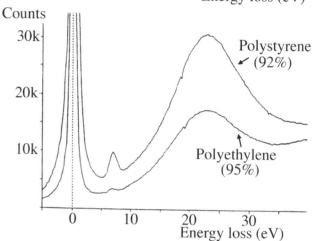

Figure 40.12. (A) Band-gap differences evident in the low-loss spectra of a Si semiconductor and SiO$_2$ and Si$_3$N$_4$ ceramic insulators. (B) The interband transition characteristic of polystyrene, clearly visible on the rise of the plasmon peak, compared with the absence of such a transition in polyethylene.

the electron–specimen interaction. However, considerable research is going on into these aspects of EELS and these fine structure studies are the future of the technique.

40.3. ENERGY-FILTERED AND SPECTRUM IMAGING

We can select the intensity in any part of the EELS spectrum and use it to form an image, either in a digital manner by modulating the signal to the STEM screen or in an analog manner in the energy-selecting TEM. A variety of images can be formed in this way and they have several advantages over conventional TEM and STEM images. We will describe the experimental procedures first and then discuss the different types of images.

40.3.A. STEM Digital Imaging

In SEELS, the ramp voltage to the magnetic prism must be held at a constant setting, so only those electrons in the energy range accepted by the slit pass through the spectrometer. In PEELS you select the output from specific diodes. In either case you've used the spectrometer to select electrons of a fixed energy range. If these electrons are then allowed to hit a detector, the signal can be used to form energy-filtered images. In a TEM/STEM you use the signal from the EELS scintillator to modulate the STEM CRT, while in a dedicated STEM the BF detector sits beyond the EELS and so all your BF images are energy-filtered. To avoid image shifts due to scanning of the beam on the specimen, you must descan the beam using a set of post-specimen coils, which are usually present in the TEM or STEM column as a matter of course. In SEELS, if the spectrometer slit width is too large, your image will suffer from chromatic aberration because electrons of different energy are focused at different planes. In PEELS, the intensity is controlled by the total spectral-acquisition time per pixel. Here, an FEG is best if high-resolution images with reasonable pixel numbers are to be acquired.

> With PEELS, it is possible to collect a spectrum in a sufficiently short time (< 50 ms) that you can create images in which a complete spectrum is stored at each pixel and all data processing is carried out after the acquisition.

Then we have what is known as a *spectrum image* (Jeanguillaume and Colliex 1989). Such images contain immense amounts of data and you need sophisticated hardware and software routines to handle the data. For example, a 512×512 pixel image with a full 1024 channel spectrum at each pixel contains more than 25 Mb of data. Because of the relatively long time to acquire spectrum images, drift correction and other PEELS corrections are necessary (Hunt and Williams 1991). Within such an image you have a *complete* record of the electron–specimen interaction and from such you can create multiple images, as shown in Figure 40.13. This figure is from an Al-Li alloy and shows the distribution of the component elements. There are three strong advantages to this approach:

- ■ You can analyze the "specimen" at a later time, without putting it back in the microscope, and look for elements that were not initially thought to be present or to be important.
- ■ You can process the data in several different ways to compare quantification schemes and the possibility of discerning unexpected correlations between elemental distributions.
- ■ All the information in the EELS spectrum can be mapped discretely, creating, for example, not just elemental images, but dielectric-constant images, valence-state images, thickness images, etc.

40.3.B. TEM Analog Imaging

For analog imaging in a TEM, an Ω filter spectrometer sits between the first and second pairs of projector lenses, as shown back in Figure 37.11 for the LEO EM912. To select

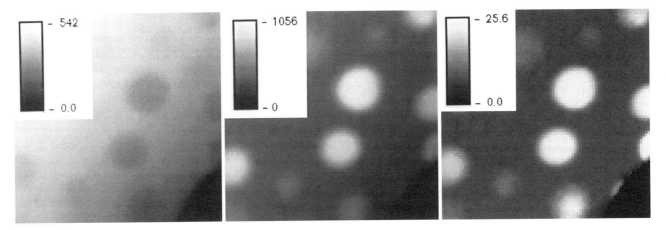

Figure 40.13. Three processed spectrum images of an Al-Li alloy aged to give a dispersion of δ' precipitates. The left image shows the absolute concentration of Al (atoms/nm²) obtained from quantification of the Al $L_{2,3}$ edge, the middle image is the absolute Li content (atoms/nm²) from the Li K edge, and the right image is the Li content (at. %) obtained from the shift of the first plasmon peak. The inserts show the correlation between image intensity and the range of composition imaged.

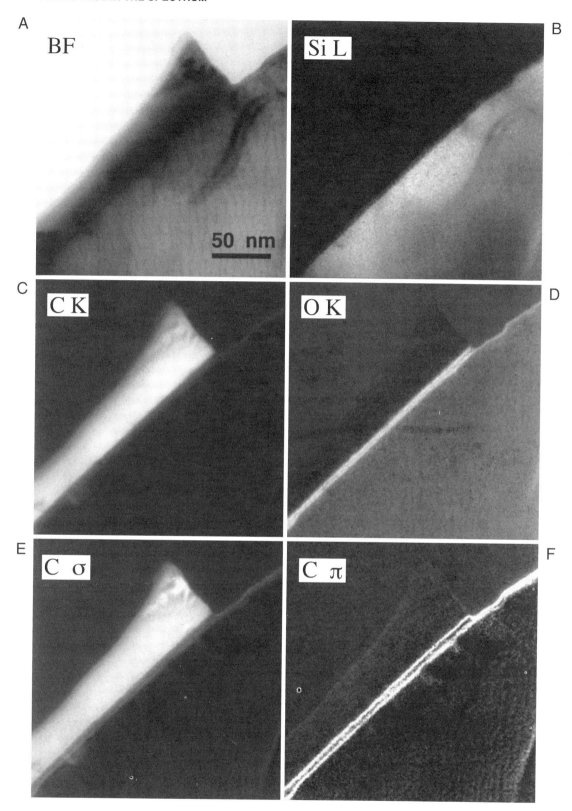

Figure 40.14. (A) TEM BF and (B–F) a series of electron spectroscopic images revealing the Si, C, and O elemental distributions and the carbon bonding maps at the interface between a diamond-like carbon film and a Si substrate. In the oxygen-rich amorphous layer at the interface the carbon atoms exhibit a double layer of π bonds while the carbon film itself is predominantly σ bonded, indicating a high degree of diamond-like character.

the electrons for this imaging (ESI), you shift the spectrum relative to the slit that is positioned after the filter but before the final projector lens. In the LEO instrument, you make the shift by increasing the accelerating voltage of the microscope by $+E$ in order to keep the energy-loss electrons of interest $(-E)$ on the optic axis; this shift correction is prealigned for the chosen kV. You can filter either an image or DP simply by changing the strength of the intermediate lens preceding the Ω filter. In addition, you can also see the energy-loss spectrum on the TEM screen. This method of spectral display has the advantage that the angular distribution of the energy-loss electrons is spatially resolved, although the absolute intensity has to be determined by digitizing the spectrum or using a microdensitometer. However, this is not the major mode of operation of the instrument, which is optimized for electron spectroscopic imaging (ESI), and many examples are given in a special issue of the *Journal of Microscopy* (Knowles 1994).

For ESI, you adjust the energy window by varying the slit width. With a 20-eV window you can obtain images with a chromatic-aberration limit of ~2.5 nm, which compares well with normal TEM C_c-limited resolution. Resolution may be as good as 0.5 nm under ideal conditions. You can select the area to image via a selected-area aperture, or by using Kohler illumination conditions, in which a small parallel beam of electrons is created.

> A drawback to this ESI process is that, while background-subtracted core-loss images are easily obtainable, they are not quantitative if significant changes in specimen thickness occur.

However, the images can be acquired in a few seconds, rather than many minutes or hours for a digital image, and a range of filtered images are compared with a conventional TEM BF image in Figure 40.14A–F. ESI is equally applicable to diffraction patterns; energy-filtered CBED is a very powerful technique for extracting more data from conventional CBED patterns, as shown in Figure 40.15. Deininger *et al.* (1994) have demonstrated how energy-filtered CBED patterns can be used to determine structure factors, lattice strains, and the accelerating voltage of the TEM. Removing the inelastic electrons removes much of the diffuse scattering from your diffraction patterns, making comparison of experimental and simulated patterns much easier. In addition to energy-filtered CBED patterns, as shown in Figure 40.15, SAD patterns can be similarly sharpened up, and used for RDF determination, as we already mentioned in Section 40.1.B

In summary, you can perform EELS imaging in two very different ways in a TEM and STEM. You can obtain a variety of images, depending on which portion of your

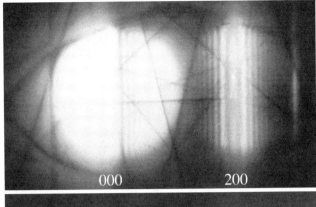

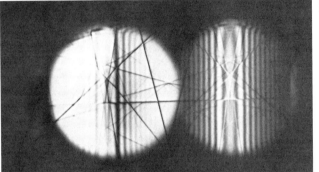

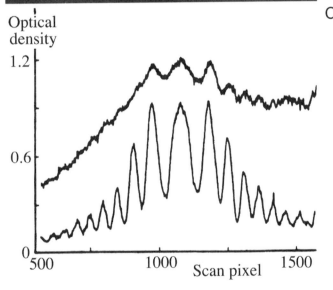

Figure 40.15. (A) Experimental CBED two-beam pattern (000 and 220) from a Si specimen, 270 nm thick. (B) The same pattern energy-filtered using the Zeiss Ω filter with an 8-eV window revealing the K-M fringes useful for thickness determination. (C) Densitometer traces across the 220 diffraction disk, unfiltered (above) and filtered (below).

spectrum is selected and the nature of your specimen. The strong forward-scattered EELS signal, combined with close to 100% detection efficiency, means that EELS imaging is much more statistically viable than thin-film X-ray mapping. This fact, combined with the enormous number of signals available in the EELS spectrum, make EELS imaging an extremely attractive technique.

Combination of energy filtering with conventional TEM imaging and diffraction will increase as the development of digital TEM technology continues. It is likely that all images will routinely be filtered to remove chromatic aberration effects, which will be a tremendous aid to the materials scientist struggling to make thin specimens from complex multiphase materials. As digital storage becomes cheaper, the ability to save complete spectrum images of all your specimens will become the norm, thus enhancing the claim that the TEM is *the* most versatile instrument for the characterization of materials.

As a final word, never forget to combine techniques wherever possible to characterize your material. If you are creative, you can even do simultaneous experiments, e.g., by constructing an STM in a TEM (Spence *et al.* 1990). We encourage you to experiment with the microscope at *all* opportunities. Don't think there is nothing new to discover; there is still ample room for you to exercise your imagination and innovation, and the TEM is a fascinating place in which to do just that.

CHAPTER SUMMARY

Both the experimental techniques and the theoretical understanding of EELS are still developing. We have introduced several specialized topics:

- Energy-loss near-edge structure (ELNES).
- Extended energy-loss fine structure (EXELFS).
- Low-loss fine structure.
- Angle-resolved (momentum transfer) EELS.
- Electron spectroscopic imaging (ESI) and spectrum imaging.

However, we have only given you a suspicion of the potential of these topics. If EELS becomes a technique you use in your research, or if you have the time, we recommend watching developments of the technique in the journals referenced at the end of this chapter. EELS, particularly fine structure and imaging, is one of the most dynamic areas of TEM development.

REFERENCES

General References

Egerton, R.F. (1996) *Electron Energy-Loss Spectroscopy in the Electron Microscope,* 2nd edition. Plenum Press, New York.
Raether, H. (1965) *Electron Energy-Loss Spectroscopy, Springer Tracts in Modern Physics,* Springer-Verlag, New York.
Teo, B.K. and Joy, D.C. (1981) *EXAFS Spectroscopy; Techniques and Applications,* Plenum Press, New York.

Specific References

Batson, P.E. (1995) *Ultramicroscopy* **59**, 63.
Browning, N.D. and Pennycook, S.J. (1995) *J. Microsc.* **180**, 230.
Bruley, J., Williams, D.B., Cuomo, J.J., and Pappas, D.P. (1995) *J. Microsc.* **180**, 22.
Brydson, R. (1991) *EMSA Bulletin* **21**, 57.

Cockayne, D.J.H., McKenzie, D., and Muller, D. (1991) *Microsc. Microanal. Microstruct.* **2**, 359.
Deininger, C., Necker, G., and Mayer, J. (1994) *Ultramicroscopy* **54**, 15.
Hunt, J.A. (1995) in *Microbeam Analysis-1995* (Ed. E.S. Etz), p. 215, VCH Publishers, New York.
Hunt, J.A. and Williams, D.B. (1991) *Ultramicroscopy* **38**, 47.
Jeanguillaume, C. and Colliex, C. (1989) *Ultramicroscopy* **28**, 252.
Knowles, K.M., Ed. (1994) *J. Microsc.* **174**, 131.
Leapman, R.D. and Silcox, J. (1979) *Phys. Rev. Lett.* **42**, 1362.
Leapman, R.D., Grunes, L.A., and Fejes, P.L. (1982) *Phys. Rev.* **B26**, 614.
Qian, M., Sarikaya, M., and Stern, E.A. (1995) *Ultramicroscopy* **59**, 137.
Schattschneider, P. and Exner, A. (1995) *Ultramicroscopy* **59**, 241.
Sklad, P., Angelini, P., and Sevely, J. (1992) *Phil Mag.* **A65**, 1445.
Spence, J.C.H., Lo, W., and Kuwabara, M. (1990) *Ultramicroscopy* **33**, 69.
Wang, Y.Y., Cheng, S.C., Dravid, V.P., and Zhang, F.C. (1995) *Ultramicroscopy* **59**, 109.
Williams, D.B. and Edington, J.W. (1976) *J. Microsc.* **108**, 113.

Index

A$_3$B ordered fcc, 245
Aberration, 97–98, 137, 352, 544; *see also*
    Chromatic aberration; Spherical
    aberration
  aberration-free focus, 468
  coma, 469, 473
  function, 460
Absorption
  of electrons, 223, 374
    anomalous, 298, 375
    contrast 354
    distance, 397
    parameters, 395
  of X-rays, 589
    absorption-free intensity ratio, 615
    correction, 610, 612, 613, 630
    edge, 604, 659
    extrapolation techniques for correction,
      614
    path length, 614
Accelerating voltage
  calibration of, 151
  continuous kV control for CBED, 338
  effect on Bloch waves, 217
  effect on EELS, 669, 702
  effect on Ewald sphere, 199
  effect on X-rays, 578, 583, 589, 609, 624
Adaptive filter, 523–524
Airy disk, 28, 100
ALCHEMI, 616
Allowed reflections, 272
Amorphous
  carbon, 354, 504, 582, 671; *see also* Holey
    carbon film
  germanium, 504, 511
  layer, 522
  materials, 274
  specimen, 353
Amplitude contrast, 349, 351–360, 372; *see*
    also Contrast

Amplitude of diffracted beam, 203, 207, 239,
    254, 405
Amplitude-phase diagrams, 430
Analog
  collection, 644
  to digital converter, 645
  images, 107
  pulse processing, 563
Analytical electron microscopy, 555–703
Angle, 30; *see also* Bragg; Collection
    semiangle; Convergence semiangle;
    Incidence semiangle
Angle-resolved EELS, 698
Angular-momentum quantum number, 691
Annular dark field (ADF), 352, 358, 364; *see*
    also Dark field
  detector
  image, 145
Anodic dissolution, 161
Anomalous
  X-ray generation, 601; *see also* Absorption
Anticontaminator, 121, 124
Antiphase (domain) boundaries, 245, 381, 389,
    478
Aperture, 21, 85–104, 502, 507, 524; *see also*
    Diaphragm
  condenser (C2), 78, 133, 135, 139, 305, 306,
    308, 542, 589, 683
    alignment of C2, 137
    virtual C2, 139
  differential pumping, 123, 641
  function, 460, 463
  objective, 103, 142, 351, 355, 358, 363, 373,
    425, 441, 486, 490, 502–503, 534, 647
  virtual, 466, 579
Artifact
  in EELS, 671
  in image, 10, 11, 494, 504, 512, 534
  of specimen preparation, 163, 170
  X-ray peak, 578, 582, 590, 591

Artificial color, 507
Artificial superlattice, 246
Ashby–Brown contrast, 417
Astigmatism, 98
  condenser, 138
  intermediate, 148
  objective, 147, 188, 463, 469, 473
Atomic
  basis, 242
  number, 39
    correction factor, 610
  scattering amplitude, 43, 240, 243, 276, 538,
    629
  scattering factor, 41, 239, 360
  structure, 441
Auger electron, 52, 658, 692
  spectrometer, 52
Automatic beam alignment, 511
Automatic peak identification, 590
Averaging images, 506
Axis-angle pair, 284, 285

Back focal plane, 90, 140, 186, 188, 307, 641;
    see also Lens
Background, *see also* Bremsstrahlung
  extrapolation, 673
  modeling, 602
  subtraction, 507, 601, 604, 609, 674
Backscattered electron (BSE), 23, 38, 211, 542,
    575, 576
  detection, 113
Baking, 123
Band gap (semiconductor), 195, 219, 229, 233,
    234, 236, 559
  image, 698
Bandwidth, 109, 110
Bar, 119
Barn, 24
Basal plane, 410
Basis vectors, 515–516

Note: all materials examples are indexed under "Materials examples in text."

Acknowledgments for Reproduction of Figures

TEM is a visual science, and any TEM text is heavily dependent on figures and halftones to transmit its message. We have been fortunate to work with many colleagues over the years who have generously given us fine examples of the art and science of TEM; we would like to acknowledge them here. We have also used the work of others, whose permission has been sought as listed below.

Chapter 1

Figure 1.1: From Ruska, E. (1980) *The Early History of the Electron Microscope*, Fig. 6, reproduced by permission of S. Herzel Verlag GmbH & Co.
Figures 1.4A: Courtesy of S.M. Zemyan.
Figures 1.4B,C: Courtesy of S.M. Merchant.
Figure 1.6: Courtesy K.S. Vecchio.
Figure 1.7: Courtesy T. Hayes, from Hayes, T. (1980) in Johari, O. (Ed.) SEM—1980 **1**, 1, Fig. 8, reproduced by permission of Scanning Microscopy International.
Figure 1.9A: Courtesy of JEOL USA Inc.
Figure 1.9B: Courtesy of NSA Hitachi Scientific Instruments, Ltd.
Figure 1.9C: Courtesy of Philips Electronic Instruments, Inc.
Figure 1.9D: Courtesy of V.G. Scientific.

Chapter 2

Figure 2.4: Courtesy of J. Bruley and V.J. Keast.
Figure 2.8: Modified from Hecht, E. (1988) *Optics* Fig. 10.21, Addison-Wesley.
Figure 2.11A,D: Courtesy of K.S. Vecchio.
Figure 2.11C: Courtesy of D.W. Ackland.

Chapter 3

Figure 3.3: Courtesy of D.E. Newbury, from Newbury, D.E. (1986) in Joy, D.C., *et al.* (Eds.) *Principles of Analytical*

Electron Microscopy, p. 6, Fig. 2, reproduced by permission of Plenum Press.
Figure 3.4: Courtesy of D.E. Newbury, from data in Newbury D.E. (1986) in Joy, D.C., *et al.* (Eds.) *Principles of Analytical Electron Microscopy*, p. 8, Table II, reproduced by permission of Plenum Press.

Chapter 4

Figure 4.1: Courtesy of D.E. Newbury, from Newbury, D.E. (1986) in Joy, D.C., *et al.* (Eds.) *Principles of Analytical Electron Microscopy*, p. 20, Fig. 4, reproduced by permission of Plenum Press.
Figure 4.3: From Woldseth, R. (1973) *X-ray Energy Spectrometry*, Fig. 3, reproduced by permission of Kevex Instruments.
Figure 4.4: From Williams, D.B. (1987) *Practical Analytical Electron Microscopy in Materials Science*, 2nd Edition, Fig. 4.3, reproduced by permission of Philips Electron Optics.
Figure 4.11: Courtesy of L.W. Hobbs, from Hobbs, L.W. (1979) in Hren, J.J., *et al.* (Eds.) *Introduction to Analytical Electron Microscopy*, Fig. 17.2, reproduced by permission of Plenum Press.
Figure 4.12: Courtesy of L.W. Hobbs, from Hobbs, L.W. (1979) in Hren, J.J., *et al.* (Eds.) *Introduction to Analytical Electron Microscopy*, Fig. 17.4, reproduced by permission of Plenum Press.
Table 4.1: Courtesy of J.I. Goldstein, from Goldstein, J.I., *et al.* (1992) *Scanning Electron Microscopy and X-ray Microanalysis*, 2nd edition, Table 3.11, reproduced by permission of Plenum Press.
Table 4.2: Courtesy of N.J. Zaluzec and J.F. Mansfield, from Zaluzec, N.J. and Mansfield, J.F. in Rajan, K. (Ed.) *Intermediate Voltage Electron Microscopy and Its Ap-*

plication to Materials Science (1987), p. 29, Table 1, reproduced by permission of Philips Electron Optics.

Chapter 5

Figure 5.1: Modified from Hall, C.E. (1966) *Introduction to Electron Microscopy*, Fig. 7.8, McGraw-Hill.

Figure 5.4B: Courtesy of J.I. Goldstein, from Goldstein, J.I., *et al.* (1992) *Scanning Electron Microscopy and X-ray Microanalysis*, 2nd edition, Fig. 2.7, reproduced by permission of Plenum Press.

Figure 5.5: Courtesy of D.W. Ackland.

Figure 5.6: Courtesy of D.W. Ackland.

Figure 5.7A: Modified from Crewe, A.V., *et al.* (1969) *Rev. Sci. Instrum.* **40**, 241, Fig. 2.

Figure 5.7B: Courtesy of D.W. Ackland.

Figure 5.10: Courtesy of J.R. Michael, from Michael, J.R. and Williams, D.B. (1987) *J. Microsc.* **147**, 289, Fig. 3, reproduced by permission of the Royal Microscopical Society.

Figure 5.11A: Modified from Williams, D.B. (1987) *Practical Analytical Electron Microscopy in Materials Science,* 2nd Edition, Fig. 2.12b, Philips Electron Optics.

Figure 5.12: Courtesy of J.R. Michael, from Michael, J.R. and Williams, D.B. (1987) *J. Microsc.* **147**, 289, Fig. 2, reproduced by permission of the Royal Microscopical Society.

Figure 5.13A: Courtesy of D.W. Ackland.

Figure 5.13B: Courtesy of NSA Hitachi Scientific Instruments Ltd.

Chapter 6

Figure 6.7: Courtesy of D.W. Ackland.

Figure 6.8A: Courtesy of Philips Electronic Instruments Inc.

Figure 6.8B: Courtesy of Kratos Ltd.

Figure 6.8C: From Mulvey, T. (1974) Electron Microscopy—1974, p. 17, Fig. 1, reproduced by permission of the Australian Academy of Science.

Figure 6.8D: From Reimer, L. (1993) *Transmission Electron Microscopy*, 3rd edition, Fig. 2.12, reproduced by permission of Springer Verlag.

Figure 6.9: Modified from Reimer, L. (1993) *Transmission Electron Microscopy*, 3rd edition, Fig. 2.3, Springer Verlag.

Figure 6.10B: Courtesy of A.O. Benscoter.

Figure 6.11: Modified from Reimer, L. (1993) *Transmission Electron Microscopy*, 3rd edition, Fig. 2.13, Springer Verlag.

Figure 6.14: Modified from Reimer, L. (1993) *Transmission Electron Microscopy*, 3rd edition, Fig. 4.23, Springer Verlag.

Chapter 7

Figure 7.1: Modified from Stephen, J., *et al.* (1975) *J. Phys. E* **8**, 607, Fig. 2.

Figure 7.5: Modified from Williams, D.B. (1987) *Practical Analytical Electron Microscopy in Materials Science*, 2nd Edition, Fig. 1.2, Philips Electron Optics.

Figure 7.6: From Berger, S.D., *et al.* (1985) *Electron Microscopy and Analysis*, p. 137, Fig.1, reproduced by permission of The Institute of Physics Publishing.

Chapter 8

Figure 8.1: Courtesy of W.C. Bigelow, from Bigelow, W.C. (1994) *Vacuum Methods in Electron Microscopy*, Fig. 4.1, reproduced by permission of Portland Press, Ltd.

Figure 8.2: Courtesy of W.C. Bigelow, from Bigelow, W.C. (1994) *Vacuum Methods in Electron Microscopy*, Fig. 5.1, reproduced by permission of Portland Press, Ltd.

Figure 8.3: Courtesy of Leybold Vacuum Products Inc.

Figure 8.4: Courtesy of W.C. Bigelow, from Bigelow, W.C. (1994) *Vacuum Methods in Electron Microscopy*, Fig. 7.1, reproduced by permission of Portland Press, Ltd.

Figure 8.6: Courtesy of Gatan Inc.

Figure 8.7: From Valdrè, U., and Goringe, M.J. (1971) in Valdrè, U. (Ed.) *Electron Microscopy in Materials Science*, p. 217, Fig. 6, reproduced by permission of Academic Press Inc.

Figure 8.8: Courtesy of NSA Hitachi Scientific Instruments Ltd.

Figure 8.9A,B: Courtesy of Gatan Inc.

Figure 8.10A,B: Courtesy of Gatan Inc.

Figure 8.11: Courtesy of Gatan Inc.

Figure 8.12: From Komatsu, M., *et al.* (1994) *Journal of the American Ceramic Society* **77**, 839, Fig.1, reproduced by permission of The American Ceramic Society.

Figure 8.13: Courtesy of NSA Hitachi Scientific Instruments Ltd.

Chapter 9

Figure 9.6: Modified from Reimer, L. (1993) *Transmission Electron Microscopy*, 3rd edition, Fig. 4.14a, Springer Verlag.

Figure 9.16: From Williams, D.B. (1987) *Practical Analytical Electron Microscopy in Materials Science*, 2nd Edition, Fig. 1.7, reproduced by permission of Philips Electron Optics.

Figure 9.18B,C,D: Courtesy of D.W.Ackland.

Figure 9.19: From Edington, J.W. (1976) *Practical Electron Microscopy in Materials Science*, Fig. 1.5, reproduced by permission of Philips Electron Optics.

Figure 9.20: Courtesy of S. Ramamurthy.
Figure 9.21: Courtesy of D.W. Ackland.
Figure 9.23: Courtesy of D.W. Ackland.
Figure 9.24: Courtesy of D.W. Ackland.
Figure 9.25: Courtesy of S. Ramamurthy.
Table 9.1: From Williams, D.B. (1987) *Practical Analytical Electron Microscopy in Materials Science*, 2nd Edition, Table 2.4, reproduced by permission of Philips Electron Optics.
Table 9.2: From Williams, D.B. (1987) *Practical Analytical Electron Microscopy in Materials Science*, 2nd Edition, Table 2.2, reproduced by permission of Philips Electron Optics.

Chapter 10

Figure 10.1: Modified from Médard, L., *et al.* (1949) *Rev. Met.* **46**, 549, Fig.5.
Figure 10.2: Courtesy of SPI Inc.
Figure 10.3: Courtesy of Gatan Inc.
Figure 10.4: Courtesy of VCR Inc.
Figure 10.7A: From Thompson-Russell, K.C. and Edington, J.W. (1977) *Electron Microscope Specimen Preparation Techniques in Materials Science*, Fig. 9, reproduced by permission of Philips Electron Optics.
Fig. 10.7b: From Thompson-Russell, K.C. and Edington, J.W. (1977) *Electron Microscope Specimen Preparation Techniques in Materials Science*, Fig. 7, reproduced by permission of Philips Electron Optics.
Figure 10.8: From Thompson-Russell, K.C. and Edington, J.W. (1977) *Electron Microscope Specimen Preparation Techniques in Materials Science*, Fig. 12, reproduced by permission of Philips Electron Optics.
Figure 10.9: Modified from Thompson-Russell, K.C. and Edington, J.W. (1977) *Electron Microscope Specimen Preparation Techniques in Materials Science*, Fig. 11, Philips Electron Optics.
Figure 10.10: Courtesy of R. Alani, Gatan Inc.
Figure 10.11: Courtesy of A.G. Cullis, from Cullis, A.G., *et al.* (1985) Ultramicroscopy **17**, 203, Figs. 1a, 3, reproduced by permission of Elsevier Science B.V.
Figure 10.12: From van Hellemont, J., *et al.* (1988) in Bravman, J., *et al.* (Eds.) *Specimen Preparation for Transmission Electron Microscopy of Materials*, Mat. Res. Soc. Symp. **115**, 247, Fig. 1, reproduced by permission of MRS.
Figure 10.15: From Thompson-Russell, K.C. and Edington, J.W. (1977) *Electron Microscope Specimen Preparation Techniques in Materials Science*, Figs. 20, 21, reproduced by permission of Philips Electron Optics.
Figure 10.16: From Thompson-Russell, K.C. and Edington, J.W. (1977) *Electron Microscope Specimen Preparation*

Techniques in Materials Science, Fig. 25, reproduced by permission of Philips Electron Optics.
Figure 10.17: From Hetherington, C.J.D. (1988) in Bravman, J., *et al.* (Eds.) *Specimen Preparation for Transmission Electron Microscopy of Materials*, Mat. Res. Soc. Symp. **115**, 143, Fig. 1, reproduced by permission of MRS.
Figure 10.18: Modified from Dobisz, E.A., *et al.* (1986) *J. Vac. Sci. Technol. B* **4**, 850, Fig. 1, reproduced by permission of MRS.
Figure 10.19: From Fernandez, A. (1988) in Bravman, J., *et al.* (Eds.) *Specimen Preparation for Transmission Electron Microscopy of Materials*, Mat. Res. Soc. Symp. **115**, 119, Fig. 1, reproduced by permission of MRS.
Figure 10.20: Courtesy of P. Goodhew, from Goodhew, P.J. (1988) in Bravman, J., *et al.* (Eds.) *Specimen Preparation for Transmission Electron Microscopy of Materials*, Mat. Res. Soc. Symp. **115**, 52, reproduced by permission of MRS.
Table 10.1: Courtesy of T. Malis.

Chapter 11

Table 11.1: Modified from Hirsch, P.B., *et al.* (1977) *Electron Microscopy of Thin Crystals*, 2nd edition, p. 19, Krieger.

Chapter 13

Table 13.2: Modified from Reimer, L. (1993) *Transmission Electron Microscopy*, 3rd edition, Table 7.2, p. 296, Springer Verlag.

Chapter 14

Figure 14.2: Modified from Hashimoto, H., *et al.* (1962) *Proc. Roy. Soc. (London)* **A269**, 80, Fig. 2.
Table 14.2: Modified from Reimer, L. (1993) *Transmission Electron Microscopy*, 3rd edition, Table 3.2, p. 58, Springer Verlag.

Chapter 16

Figure 16.5: Courtesy of M.L. Jenkins, from Jenkins, M.L., *et al.* (1976) *Phil. Mag.* **34**, 1141, Fig. 2, reproduced by permission of Taylor and Francis.
Figure 16.6: Courtesy of B.C. De Cooman.
Figure 16.7: From Dodsworth, J., *et al.* (1983) *Adv. Ceram.* **6**, 102, Fig. 3, reproduced by permission of the American Ceramic Society.
Figure 16.8: Courtesy of B.C. De Cooman.
Figure 16.9: From Gajdardziska-Josifovska, M., *et al.* (1995) *Ultramicroscopy* **58**, 65, Fig. 1, reproduced by permission of Elsevier Science B.V.

Figure 16.10: Courtesy of S. McKernan.

Figure 16.11: From Hahn, T. (Ed.) *International Tables for Crystallography A*, pp. 538–539, No. 164, reproduced by permission of The International Union of Crystallography.

Table 16.1: Modified from Edington, J.W. (1976) *Practical Electron Microscopy in Materials Science*, Appendix 8, Van Nostrand Reinhold.

Chapter 17

Figure 17.2: From Edington, J.W. (1976) *Practical Electron Microscopy in Materials Science*, Fig. 2.16, reproduced by permission of Philips Electron Optics.

Figure 17.9: Modified from Hirsch, P.B., *et al.* (1977) *Electron Microscopy of Thin Crystals*, 2nd edition, Fig. 4.11, Krieger.

Figure 17.7: From Carter, C.B., *et al.* (1981) *Phil. Mag.* **A43**, 441, Fig. 5c, reproduced by permission of Taylor and Francis.

Figure 17.10: From Driver, J.H., *et al.* (1972) *Phil Mag.* **26**, 1227, Fig. 3, reproduced by permission of Taylor and Francis.

Figure 17.11A–C: From Lewis, M.H., and Billingham, J. (1972) *JEOL News* **10e**(1), 8, Fig. 3, reproduced by permission of JEOL USA Inc.

Figure 17.11D: Modified from Sauvage, M. and Parthè, E. (1972) *Acta Cryst.* **A28**, 607, Fig. 2.

Figure 17.12: Modified from Carter, C.B., *et al.* (1981) *Phil. Mag.* **A43**, 441, Figs. 5a, b.

Figure 17.13: Modified from Carter, C.B., *et al.* (1980) *J. Electron Microsc.* **63**, 623, Fig. 8.

Figure 17.14: Modified from Carter, C.B. (1984) *Phil. Mag.* **A50**, 133, Figs. 1–3.

Chapter 18

Figure 18.2: From Edington, J.W. (1976) *Practical Electron Microscopy in Materials Science*, Fig. A1.7, reproduced by permission of Philips Electron Optics.

Figure 18.7: Courtesy of S. Ramamurthy.

Figure 18.9: Courtesy of S. McKernan.

Figure 18.10A,C: Courtesy of S. McKernan.

Figure 18.10B,D,E: From Vainshtein, B.K., *et al.* (1992) in Cowley, J.M. (Ed.) *Electron Diffraction Techniques* **1**, Fig. 6.13, reproduced by permission of Oxford University Press.

Figure 18.11: Modified from James, R.W. (1965) in Bragg, L. (Ed.) *The Optical Principles of the Diffraction of X-ray*, Vol. II of the Crystalline State, Figs. 170, 184, Cornell University Press.

Figure 18.12: Courtesy of D.J.H. Cockayne, from Sproul, A.,

et al. (1986) *Phil. Mag.* **B54**, 113, Fig. 1, reproduced by permisson of Taylor and Francis.

Figure 18.13: From Graczyk, J.F. and Chaudhari, P. (1973) *Phys. stat. sol.* (b), **58**, 163, Fig. 10a, reproduced by permission of Akademie Verlag GmbH.

Figure 18.14: Courtesy of A. Howie, from Howie, A. (1988) in Buseck, P.R., *et al.* (Eds.) *High-Resolution Transmission Microscopy and Associated Techniques*, p. 607, Fig. 14.12, reproduced by permission of Oxford University Press.

Figure 18.15: From Tietz, L.A., *et al.* (1995) *Ultramicroscopy* **60**, 241, Figs. 2, 3, 4, reproduced by permision of Elsevier Science B.V.

Figure 18.16: From Tietz, L.A., *et al.* (1995) *Ultramicroscopy* **60**, 241, Fig. 5, reproduced by permision of Elsevier Science B.V.

Figure 18.17: From Andrews, K.W., *et al.* (1971) *Interpretation of Electron Diffraction Patterns*, 2nd edition, Fig. 41, reproduced by permission of Plenum Press.

Figure 18.18: From Andrews, K.W., *et al.* (1971) *Interpretation of Electron Diffraction Patterns*, 2nd edition, Fig. 41, reproduced by permission of Plenum Press.

Figure 18.19: From Andrews, K.W., *et al.* (1971) *Interpretation of Electron Diffraction Patterns*, 2nd edition, Fig. 41, reproduced by permission of Plenum Press.

Figure 18.20: From Edington, J.W. (1976) *Practical Electron Microscopy in Materials Science*, Fig. 2.20, reproduced by permission of Philips Electron Optics.

Chapter 19

Figure 19.6A: Courtesy of G. Thomas, from Levine, E., *et al.* (1966) *J. Appl. Phys.* **37**, 2141, Fig. 1a, reproduced by permission of the American Institute of Physics.

Figure 19.7: Modified from Okamoto, P.R., *et al.* (1967) *J. Appl. Phys.* **38**, 289, Fig. 5.

Figure 19.8: Courtesy of S. Ramamurthy.

Figure 19.9A: Modified from Thomas, G. and Goringe, M.J. (1979) *Transmission Electron Microscopy of Metals*, Fig. 2.30, John Wiley & Sons Inc.

Figure 19.9B: Modified from Edington, J.W. (1976) *Practical Electron Microscopy in Materials Science*, Fig. 2.27, Van Nostrand Reinhold.

Figure 19.11: Modified from Thomas, G. and Goringe, M.J. (1979) *Transmission Electron Microscopy of Metals*, Fig. 2.29, John Wiley & Sons Inc.

Chapter 20

Figure 20.2A: Courtesy of K.S. Vecchio, from Williams, D.B., *et al.* (Eds.) (1992) *Images of Materials*, Fig. 6.5, reproduced by permission of Oxford University Press.

Figure 20.2B: Courtesy of K.S. Vecchio, from Williams, D.B., *et al.* (Eds.) (1992) *Images of Materials*, Fig. 6.17, reproduced by permission of Oxford University Press.

Figure 20.3: Modified from Williams, D.B. (1987) *Practical Analytical Electron Microscopy in Materials Science*, 2nd Edition, Fig. 6.6, Philips Electron Optics.

Figure 20.5: From Lyman, C.E., *et al.* (Eds.) (1990) *Scanning Electron Microscopy, X-ray Microanalysis and Analytical Electron Microscopy—a Laboratory Workbook*, Fig. A27.2, reproduced by permission of Plenum Press.

Figure 20.6: Courtesy of J.F. Mansfield, from Mansfield, J.F. (1984) *Convergent Beam Diffraction of Alloy Phases*, Fig. 5.3, reproduced by permission of Institute of Physics Publishing.

Figure 20.7C: Courtesy of R. Ayer.

Figure 20.8A: Modified from Ayer, R. (1989) *J. Electron Microscopy Tech.* **13**, 3, Fig. 3.

Figure 20.9: From Williams, D.B. (1987) *Practical Analytical Electron Microscopy in Materials Science*, 2nd Edition, Fig. 6.13, reproduced by permission of Philips Electron Optics.

Figure 20.10: From Williams, D.B. (1987) *Practical Analytical Electron Microscopy in Materials Science*, 2nd Edition, Fig. 6.14, reproduced by permission of Philips Electron Optics.

Figure 20.11: Courtesy of W.A.T. Clark, from Heilman, P., *et al.* (1983) *Acta Met.* **31**, 1293, Fig. 4, reproduced by permision of Elsevier Science B.V.

Figure 20.12: From Williams, D.B. (1987) *Practical Analytical Electron Microscopy in Materials Science*, 2nd Edition, Fig. 6.9, reproduced by permission of Philips Electron Optics.

Figure 20.13: From Williams, D.B. (1987) *Practical Analytical Electron Microscopy in Materials Science*, 2nd Edition, Fig. 6.16, reproduced by permission of Philips Electron Optics.

Figure 20.14: Courtesy of C.M. Sung.

Figure 20.15: Courtesy of B. Ralph, from Williams, D.B. (1987) *Practical Analytical Electron Microscopy in Materials Science*, 2nd Edition, Fig. 6.18, reproduced by permission of Philips Electron Optics.

Figure 20.16: Courtesy of K.S. Vecchio, from Williams, D.B., *et al.* (Eds.) (1992) *Images of Materials*, Fig. 6.14, reproduced by permission of Oxford University Press.

Chapter 21

Figure 21.1: From Williams, D.B. (1987) *Practical Analytical Electron Microscopy in Materials Science*, 2nd Edition, Fig. 4.29a, reproduced by permission of Philips Electron Optics.

Figure 21.3: From Williams, D.B. (1987) *Practical Ana-lytical Electron Microscopy in Materials Science*, 2nd Edition, Figs. 4.29b, c, reproduced by permission of Philips Electron Optics.

Figure 21.4: Courtesy of R. Ayer, from Raghavan, M., *et al.* (1984) *Met. Trans.* **15A**, 783, Fig. 6, reproduced by permission of ASM International.

Figure 21.5A: Courtesy of K.S. Vecchio, from Williams, D.B., *et al.* (Eds.) (1992) *Images of Materials*, Fig. 6.23, reproduced by permission of Oxford University Press.

Figure 21.5B: Courtesy of R. Ayer, from Ayer, R. (1989) *J. Electron Microsc. Tech.* **13**, 3, Fig. 7, reproduced by permission of John Wiley & Sons Inc.

Figure 21.7: Courtesy of K.S. Vecchio, from Williams, D.B., *et al.* (Eds.) (1992) *Images of Materials*, Fig. 6.19, reproduced by permission of Oxford University Press.

Figure 21.8: Courtesy of V.P. Dravid.

Figure 21.9A–D: Courtesy of J.W. Steeds, from Chapman J.N. and Craven, A.J. (Eds.) (1984) *Quantitative Electron Microscopy*, p. 68, Fig. 8, reproduced by permission of the Scottish Universities Summer School in Physics.

Figure 21.9E: Courtesy of K.S. Vecchio, from Williams, D.B., *et al.* (Eds.) (1992) *Images of Materials*, Fig. 6.25, reproduced by permission of Oxford University Press.

Figure 21.10: Courtesy of K.S. Vecchio, from Williams, D.B., *et al.* (Eds.) (1992) *Images of Materials*, Fig. 6.24, reproduced by permission of Oxford University Press.

Figure 21.11: Courtesy of K.S. Vecchio and V.P. Dravid, from Williams, D.B., *et al.* (Eds.) (1992) *Images of Materials*, Figs. 6.27, 6.28, reproduced by permission of Oxford University Press.

Figure 21.12: Courtesy of K.S. Vecchio, from Williams, D.B., *et al.* (Eds.) (1992) *Images of Materials*, Fig. 6.29, reproduced by permission of Oxford University Press.

Figure 21.13: Courtesy of K.S. Vecchio and V.P. Dravid, from Williams, D.B., *et al.* (Eds.) (1992) *Images of Materials*, Fig. 6.30, reproduced by permission of Oxford University Press.

Figure 21.14: Courtesy of R. McConville, from Williams, D.B., *et al.* (Eds.) (1992) *Images of Materials*, Fig. 6.33, reproduced by permission of Oxford University Press.

Figure 21.15: Courtesy of J.M. Cowley, from Liu, M. and Cowley, J.M. (1994) *Ultramicroscopy* **53**, 333, Figs. 1, 2, reproduced by permission of Elsevier Science B.V.

Figure 21.16: Courtesy of W.D. Riecke, from Williams, D.B. (1987) *Practical Analytical Electron Microscopy in Materials Science*, 2nd Edition, Fig. 6.2, reproduced by permission of Philips Electron Optics.

Figure 21.17: Courtesy of J.R. Michael.

Table 21.1: Data from Williams, D.B. (1987) *Practical Analytical Electron Microscopy in Materials Science*, 2nd Edition, p. 79, reproduced by permission of Philips Electron Optics.

Table 21.2: Data from Williams, D.B. (1987) *Practical Analytical Electron Microscopy in Materials Science*, 2nd Edition, p. 79, reproduced by permission of Philips Electron Optics.

Table 21.3: Courtesy of B.F. Buxton, from Buxton, B.F., *et al.* (1976) *Phil. Trans. Roy. Soc.* (London) **281**, 181, Table 2, reproduced by permission of The Royal Society.

Table 21.4: Courtesy of B.F. Buxton, from Buxton, B.F., *et al.* (1976) *Phil. Trans. Roy. Soc.* (London) **281**, 181, Table 3, reproduced by permission of The Royal Society.

Table 21.6: Courtesy of B.F. Buxton, from Buxton, B.F., *et al.* (1976) *Phil. Trans. Roy. Soc.* (London) **281**, 181, Table 4, reproduced by permission of The Royal Society.

Table 21.7: Courtesy of J.W. Steeds, from Steeds, J.W., *et al.* (1983) *J. Appl. Cryst.* **16**, 317, Tables 5, 6, reproduced by permission of the International Union of Crystallography.

Chapter 22

Figure 22.5: Courtesy of K.A. Repa.

Figure 22.6: From Williams, D.B. (1987) *Practical Analytical Electron Microscopy in Materials Science*, 2nd Edition, Fig. 3.7d, reproduced by permission of Philips Electron Optics.

Figure 22.7: Courtesy of K.B. Reuter.

Figure 22.8: From Williams, D.B. (1987) *Practical Analytical Electron Microscopy in Materials Science*, 2nd Edition, Fig. 3.7c, reproduced by permission of Philips Electron Optics.

Figure 22.9A,B: Courtesy of H. Tsai, from Williams, D.B. (1987) *Practical Analytical Electron Microscopy in Materials Science*, 2nd Edition, Fig. 1.19a, b, reproduced by permission of Philips Electron Optics.

Figure 22.9C: Courtesy of K.-R. Peters.

Figure 22.10: Modified from Williams, D.B. (1983) in Krakow, W., *et al.* (Eds.) *Electron Microscopy of Materials*, Mat. Res. Soc. Symp. **31**, 11, Figs. 3a, b, MRS.

Figure 22.11A,B: Courtesy of I.M. Watt, from Watt, I.M. (1996) *The Principles and Practice of Electron Microscopy*, 2nd edition, Figs. 5.5a, b, reproduced by permission of Cambridge University Press.

Figure 22.12: Courtesy of M.M.J. Treacy, from Williams, D.B. (1987) *Practical Analytical Electron Microscopy in Materials Science*, 2nd Edition, Fig. 5.26b, reproduced by permission of Philips Electron Optics.

Figure 22.14: Courtesy of S.J. Pennycook, from Pennycook, S.J., *et al.* (1986) *J. Microsc.* **144**, 229, Fig. 8, reproduced by permission of the Royal Microscopical Society.

Figure 22.15A,B: Courtesy of S.J. Pennycook, from Lyman, C.E. (1992) Microscopy: The Key Research Tool, special publication of the *EMSA Bulletin* **22**, 7, Fig. 7, reproduced by permission of MSA.

Figure 22.15C: Courtesy of S.J. Pennycook, from Browning, *et al.* (1995) *Interface Science* **2**, 397, Fig. 4d, reproduced by permission of Kluwer.

Figure 22.16A: From Edington, J.W. (1976) *Practical Electron Microscopy in Materials Science*, Fig. 2.34, reproduced by permission of Philips Electron Optics.

Figure 22.17: Courtesy of D. Cohen.

Chapter 23

Figure 23.1: Courtesy of S. Ramamurthy.

Figure 23.2: From Edington, J.W. (1976) *Practical Electron Microscopy in Materials Science*, Fig. 3.2a, reproduced by permission of Philips Electron Optics.

Figure 23.3B: Courtesy of D. Cohen.

Figure 23.3C: Courtesy of S. King.

Figure 23.5: Courtesy of D. Susnitzky.

Figure 23.7: Modified from Edington, J.W. (1976) *Practical Electron Microscopy in Materials Science*, Fig. 3.3, Van Nostrand Reinhold.

Figure 23.8: Modified from Edington, J.W. (1976) *Practical Electron Microscopy in Materials Science*, Figs. 3.4b, d, Van Nostrand Reinhold.

Figure 23.9: Courtesy of S. Ramamurthy.

Figure 23.10: From Hashimoto H., *et al.* (1962) *Proc. Roy. Soc.* (London) **A269**, 80, Fig. 11, reproduced by permission of The Royal Society.

Figure 23.11A: Courtesy of NSA Hitachi Scientific Instruments Ltd.

Figure 23.11B,C: Courtesy of D. Cohen.

Figure 23.12: From Edington, J.W. (1976) *Practical Electron Microscopy in Materials Science*, Fig. 3.3d, reproduced by permission of Philips Electron Optics.

Figure 23.13B,C: From De Cooman, B.C., *et al.* (1987) in J.D. Dow and I.K. Schuller (Eds.) *Interfaces, Superlattices, and Thin Films*, Mat. Res. Soc. Symp. **77**, 187, Fig. 1, reproduced by permission of MRS.

Chapter 24

Figure 24.4A–D: Courtesy of D. Cohen.

Figure 24.4 E,F: Modified from Gevers, R., *et al.* (1963) *Phys. stat. sol.* **3**, 1563, Table 3.

Figure 24.5: From Föll, H., *et al.* (1980) *Phys. stat. sol.* (a) **58**, 393, Figs. 6a, c, reproduced by permission of Akademie Verlag GmbH.

Figure 24.7A,B: From Lewis, M.H. (1966) *Phil. Mag.* **14**, 1003, Fig. 9, reproduced by permission of Taylor and Francis.

Figure 24.7C,D: Courtesy of S. Amelinckx, from Amelinckx, S. and Van Landuyt, J. (1978) in S. Amelinckx, *et al.* (Eds.) *Diffraction and Imaging Techniques in Material Science* **I**, p. 107, Figs. 3, 18, North-Holland.

Figure 24.8: From Rasmussen, D.R., *et al.* (1991) *Phys. Rev. Lett.* **66**, (20), 262, Fig. 2, reproduced by permission of The American Physical Society.

Figure 24.9: Courtesy of S. Summerfelt.

Figure 24.13: Modified from Metherell, A.J.F. (1975) in Valdrè, U. and Ruedl, E. (Eds.) *Electron Microscopy in Materials Science* **II**, 397, Fig. 13, Commission of the European Communities.

Figure 24.14: From Hashimoto, H., *et al.* (1962) *Proc. Roy. Soc.* (London) **A269**, 80, Fig. 15, reproduced by permission of The Royal Society.

Figure 24.16: Modified from Rasmussen, R., *et al.* (1991) *Phil. Mag.* **63**, 1299, Fig. 4.

Chapter 25

Figure 25.2B: Modified from Amelinckx, S. (1964) *Solid State Physics Suppl.* **6**, Fig. 76.

Figure 25.6A–C: Modified from Carter, C.B. (1980) *Phys. stat. sol.* (a) **62**, 139, Fig. 4.

Figure 25.6f: From Van Landuyt, J., *et al.* (1970) *Phys. stat. sol.* **41**, 271, Fig. 19, reproduced by permission of Akademie Verlag GmbH.

Figure 25.6G–H: Courtesy of B.C. De Cooman.

Figure 25.7: Modified from Hirsch, P.B., *et al.* (1977) *Electron Microscopy of Thin Crystals*, 2nd edition, Fig. 7.8, Krieger.

Figure 25.8: From Delavignette, P. and Amelinckx, S. (1962) *J. Nucl. Mat.* **5**, 17, Fig. 7, reproduced by permission of Elsevier Science B.V.

Figure 25.10: From Urban, K. (1971) in Koda, S. (Ed.) *The World Through the Electron Microscope*, Metallurgy **V**, p. 26, reproduced by permission of JEOL USA Inc.

Figure 25.11: Courtesy of A. Howie, from Howie, A. and Whelan, M.J. (1962) *Proc. Roy. Soc.* (London) **A267**, 206, Fig. 14, reproduced by permission of The Royal Society.

Figure 25.12: Modified from M. Wilkens (1978) in Amelinckx, S., *et al.* (Eds.) *Diffraction and Imaging Techniques in Material Science* **I**, p. 185, Fig. 4, North-Holland.

Figure 25.14: From Dupouy G. and Perrier, F. (1971) in Koda, S. (Ed.) *The World Through the Electron Microscope*, Metallurgy **V**, p. 100, reproduced by permission of JEOL USA Inc.

Figure 25.15A: From Modeer, B. and Lagneborg, R. (1971) in Koda, S. (Ed.) *The World Through the Electron Microscope*, Metallurgy **V**, p. 44, reproduced by permission of JEOL USA Inc.

Figure 25.15B: Courtesy of D.A. Hughes, from Hansen, N. and Hughes, D.A. (1995) *Phys. stat. sol.* (a) **149**, 155,

Fig. 5, reproduced by permission of Akademie Verlag GmbH.

Figure 25.16A: From Siems, F., *et al.* (1962) *Phys. stat. sol.* **2**, 421, Fig. 5a, reproduced by permission of Akademie Verlag GmbH.

Figure 25.16C: From Siems, F., *et al.* (1962) *Phys. stat. sol.* **2**, 421, Fig. 15a, reproduced by permission of Akademie Verlag GmbH.

Figure 25.17A: Modified from Whelan, M.J. (1958–59) *J. Inst. Met.* **87**, 392, Fig. 25a.

Figure 25.17B: Courtesy of K. Ostyn.

Figure 25.18: From Takayanagi, L. (1988) *Surface Science* **205**, 637, Fig. 5, reproduced by permission of Elsevier Science B.V.

Figure 25.19A: From Tunstall, W.J., *et al.* (1964) *Phil. Mag.* **9**, 99, Fig. 9, reproduced by permission of Taylor and Francis.

Figure 25.19B: From Amelinckx, S. in Merli, P.G. and Antisari, V.M. (Eds.) *Electron Microscopy in Materials Science*, p. 128, Fig. 45, reproduced by permission of World Scientific.

Figure 25.20: Courtesy of W. Skrotski.

Figure 25.21: Courtesy of W. Skrotski.

Figure 25.22: From Carter, C.B., *et al.* (1986) *Phil. Mag.* **A55**, 21, Fig. 2, reproduced by permission of Taylor and Francis.

Figure 25.23: From Carter, C.B., *et al.* (1981) *Phil. Mag.* **A43**, 441, Fig. 3, reproduced by permission of Taylor and Francis.

Figure 25.24: Courtesy of K. Ostyn.

Figure 25.25: Courtesy of L. Tietz.

Figure 25.26A: Courtesy of L.M. Brown, from Ashby, M.F. and Brown, L.M. (1963) *Phil. Mag.* **8**, 1083, Fig. 10, reproduced by permission of Taylor and Francis.

Figure 25.26B: Modified from Whelan, M.J. (1978) in Amelinckx, S., *et al.* (Eds.) *Diffraction and Imaging Techniques in Material Science* **I**, p. 43, Fig. 36, North-Holland.

Figure 25.26C: Courtesy of L.M. Brown, from Ashby, M.F. and Brown, L.M. (1963) *Phil. Mag.* **8**, 1083, Fig. 12, reproduced by permission of Taylor and Francis.

Figure 25.27: From Rasmussen, D.R. and Carter, C.B. (1991) *J. Electron Microsc. Technique* **18**, 429, Fig. 2, reproduced by permission of John Wiley & Sons Inc.

Figure 25.27: From Rasmussen, D.R. and Carter, C.B. (1991) *J. Electron Microsc. Technique* **18**, 429, Fig. 2, reproduced by permission of John Wiley & Sons Inc.

Chapter 26

Figure 26.7: Courtesy of S. King.

Figure 26.10: Courtesy of D.J.H. Cockayne, from Cockayne,

D.J.H. (1972) *Z. Naturforschung* **27a**, 452, Fig. 6c, reproduced by permission of Verlag der Zeitschrift für Naturforschung, Tübingen.

Figure 26.13: Modified from Carter, C.B., *et al.* (1986) *Phil. Mag.* **A55**, 1, Fig. 9.

Figure 26.15: Modified from Föll, H., *et al.* (1980) *Phys. stat. sol.* (a) **58**, 393, Figs. 6b, c.

Figure 26.17: From Heidenreich, R.D. (1964) *Fundamentals of Transmission Electron Microscopy*, Fig. 9.20, reproduced by permission of John Wiley & Sons Inc.

Figure 26.18: Courtesy of D.J.H. Cockayne, from Ray, I.L.F. and Cockayne, D.J.H. (1971) *Proc. Roy. Soc.* (London) **A325**, 543, Fig. 10, reproduced by permission of The Royal Society.

Figure 26.23: Modified from Carter, C.B. (1979) J. Phys. (A) **54** (1) 395 Fig. 8a.

Chapter 27

Figure 27.3A: From Izui, K.J., *et al.* (1977) *J. Electron Microsc.* **26**, 129, Fig. 1. reproduced by permission of the Japanese Society of Electron Microscopy.

Figure 27.3C: Courtesy of J.C.H. Spence, from Spence, J.C.H. *Experimental High-Resolution Electron Microscopy*, Fig. 5.15, reproduced by permission of Oxford University Press.

Figure 27.4B: Courtesy of J.L. Hutchison, from Hutchison, J.L., *et al.* (1991) in Heydenreich, J. and Neumann, W. (Eds.) *High-Resolution Electron Microscopy—Fundamentals and Applications*, p. 205, Fig. 3, reproduced by permission of Halle/Saale.

Figure 27.4C: Courtesy of S. McKernan.

Figure 27.4D: From Carter, C.B., *et al.* (1989) *Phil. Mag.* **A63**, 279, Fig. 3, reproduced by permission of Taylor and Francis.

Figure 27.8: From Tietz, L.A., *et al.* (1992) *Phil. Mag.* **A65**, 439, Figs. 3a, 12a, c, reproduced by permission of Taylor and Francis.

Figure 27.10: Courtesy of J. Zhu.

Figure 27.12: Modified from Vincent, R. (1969) *Phil. Mag.* **19**, 1127, Fig. 4.

Figure 27.13: Modified from Norton, M.G. and Carter, C.B. (1995) *J. Mat. Sci.* **30**, Fig. 6.

Figure 27.14: Courtesy of U. Dahmen, from Hetherington, C.J.D. and Dahmen, U. (1992) in Hawkes, P.W. (Ed.) Signal and Image Processing in Microscopy and Microanalysis, *Scanning Microscopy* Supplement **6**, 405, Fig. 9, reproduced by permission of Scanning Microscopy International.

Figure 27.15: From Heidenreich, R.D. (1964) *Fundamentals of Transmission Electron Microscopy*, Figs. 5.4, 5.6, reproduced by permission of John Wiley & Sons Inc.

Figure 27.16A: From Heidenreich, R.D. (1964) *Fundamentals of Transmission Electron Microscopy*, Fig. 11.2, reproduced by permission of John Wiley & Sons Inc.

Figure 27.16B: From Boersch, H., *et al.* (1962) *Z. Phys.* **167**, 72, Fig. 4, reproduced by permission of Springer-Verlag.

Figure 27.17: Courtesy of M. Rühle.

Figure 27.18: Modified from Kouh, Y.M., *et al.* (1986) *J. Mat. Sci.* **21**, 2689, Fig. 9.

Figure 27.19: Courtesy of M. Rühle, from Rühle, M. and Sass, S.L. (1984) *Phil. Mag.* **A49**, 759, Fig. 2, reproduced by permission of Taylor and Francis.

Figure 27.20B–E: From Carter, C.B., *et al.* (1986), *Phil. Mag.* **A55**, 21, Fig. 11, reproduced by permission of Taylor and Francis.

Chapter 28

Figure 28.4: Courtesy of R. Gronsky, from Gronsky, R. (1992) in Williams, D.B., *et al.* (Eds.) *Images of Materials*, Fig. 7.6, reproduced by permission of Oxford University Press.

Figure 28.5: Courtesy of S. McKernan.

Figure 28.6: Courtesy of S. McKernan.

Figure 28.7: Modified from Cowley, J.M. (1988) in Buseck, P.R., *et al.* (Eds.) *High-Resolution Electron Microscopy and Associated Techniques*, Fig. 1.9, Oxford University Press.

Figure 28.8: Courtesy of J.C.H. Spence, from Spence, J.C.H. (1988) *Experimental High-Resolution Electron Microscopy*, 2nd Ed., Fig. 4.3, reproduced by permission of Oxford University Press.

Figure 28.9: Modified from Rose, H. (1991) in Heydenreich, J. and Neumann, W. (Eds.) *High-Resolution Electron Microscopy—Fundamentals and Applications*, p. 6, Fig. 3, Halle/Saale.

Figure 28.10: From de Jong, A.F. and Van Dyck, D. (1993) *Ultramicroscopy* **49**, 66, Fig. 1, reproduced by permission of Elsevier Science B.V.

Figure 28.11: Courtesy of M.T. Otten, from Otten, M.T. and Coene, W.M.J. (1993) *Ultramicroscopy* **48**, 77, Fig. 8, reproduced by permission of Elsevier Science B.V.

Figure 28.12: Courtesy of M.T. Otten, from Otten, M.T. and Coene, W.M.J. (1993) *Ultramicroscopy* **48**, 77, Fig. 11, reproduced by permission of Elsevier Science B.V.

Figure 28.13: Courtesy of M.T. Otten, from Otten, M.T. and Coene, W.M.J. (1993) *Ultramicroscopy* **48**, 77, Fig. 10, reproduced by permission of Elsevier Science B.V.

Figure 28.14A,B: From Amelinckx, S., *et al.* (1993) *Ultramicroscopy* **51**, 90, Fig. 2, reproduced by permission of Elsevier Science B.V.

Figure 28.15: From Amelinckx, S., *et al.* (1993) *Ultramicros-

copy **51**, 90, Fig. 3, reproduced by permission of Elsevier Science B.V.

Figure 28.16: From Rasmussen, D.R., *et al.* (1995) *J. Microsc.* **179**, 77, Figs. 2c, d, reproduced by permission of the Royal Microscopical Society.

Figure 28.18A: Courtesy of S. McKernan.

Figure 28.18B: From Berger, A., *et al.* (1994) Ultramicroscopy **55**, 101, Fig. 4b, reproduced by permission of Elsevier Science B.V.

Figure 28.18C: Courtesy of S. Summerfelt.

Figure 28.18D: Courtesy of S. McKernan.

Figure 28.19: Courtesy of D.J. Smith.

Figure 28.21B: From Van Landuyt, J., *et al.* (1991) in Heydenreich, J. and Neumann, W. (Eds.) *High-Resolution Electron Microscopy—Fundamentals and Applications*, p. 254, Fig. 6, reproduced by permission of Halle/Saale.

Figure 28.21D: From Van Landuyt, J., *et al.* (1991) in Heydenreich, J. and Neumann, W. (Eds.) *High-Resolution Electron Microscopy—Fundamentals and Applications*, p. 254, Fig. 8, reproduced by permission of Halle/Saale.

Figure 28.22: From Nissen H.-U. and Beeli, C. (1991) in Heydenreich, J. and Neumann, W. (Eds.) *High-Resolution Electron Microscopy—Fundamentals and Applications*, p. 272, Fig. 4, reproduced by permission of Halle/Saale.

Figure 28.23: From Nissen H.-U. and Beeli, C. (1991) in Heydenreich, J. and Neumann, W. (Eds.) *High-Resolution Electron Microscopy—Fundamentals and Applications*, p. 272, Fig. 2, reproduced by permission of Halle/Saale.

Figure 28.24: From Parsons, J.R., *et al.* (1973) *Phil. Mag.* **29**, 1359, Fig. 2, reproduced by permission of Taylor and Francis.

Table 28.1: Modified from de Jong, A.F. and Van Dyck, D. (1993) *Ultramicroscopy* **49**, 66, Table 1.

Chapter 29

Figure 29.2A,B: Courtesy of M.A. O'Keefe, from O'Keefe, M.A. and Kilaas, R. (1988) in Hawkes, P.W., *et al.* (Eds.) Image and Signal Processing in Electron Microscopy, *Scanning Microscopy* Supplement **2**, p. 225, Fig. 1, reproduced by permission of Scanning Microscopy International.

Figure 29.3: From Kambe, K. (1982) *Ultramicroscopy* **10**, 223, Figs. 1a–d, reproduced by permission of Elsevier Science B.V.

Figure 29.4: Courtesy of M.A. O'Keefe, from O'Keefe, M.A. and Kilaas, R. (1988) in Hawkes, P.W., *et al.* (Eds.) Image and Signal Processing in Electron Microscopy, *Scanning Microscopy* Supplement **2**, p. 225, Fig. 4, re-

produced by permission of Scanning Microscopy International.

Figure 29.5: Modified from Rasmussen, D.R. and Carter, C.B. (1990) *Ultramicroscopy* **32**, 337, Figs. 1 and 2.

Figure 29.8: From Beeli, C. and Horiuchi, S. (1994) *Phil. Mag.* **B70**, 215, Figs. 6a–d, reproduced by permission of Taylor and Francis.

Figure 29.9: From Beeli, C. and Horiuchi, S. (1994) *Phil. Mag.* **B70**, 215, Figs. 7a–d, reproduced by permission of Taylor and Francis.

Figure 29.10: From Beeli, C. and Horiuchi, S. (1994) *Phil. Mag.* **B70**, 215, Fig. 8, reproduced by permission of Taylor and Francis.

Figure 29.11: From Jiang, J., *et al.* (1995) *Phil. Mag. Lett.* **71**, 123, Fig. 4, reproduced by permission of Taylor and Francis.

Chapter 30

Figure 30.1: Courtesy of J. Heffelfinger.

Figure 30.2: From Rasmussen, D.R., *et al.* (1995) *J. Microsc.* **179**, 77, Fig. 1b, reproduced by permission of The Royal Microscopical Society.

Figure 30.3: From Rasmussen, D.R., *et al.* (1995), *J. Microsc.* **179**, 77, Fig. 5, reproduced by permission of The Royal Microscopical Society.

Figure 30.4: Courtesy of O.L. Krivanek, from Krivanek, O.L. (1988) in Buseck, P.R., *et al.* (Eds.) *High-Resolution Electron Microscopy and Associated Techniques*, Fig. 12.6, reproduced by permission of Oxford University Press.

Figure 30.5: Courtesy of O.L Krivanek, from Krivanek, O.L. (1988) in Buseck, P.R., *et al.* (Eds.) *High-Resolution Electron Microscopy and Associated Techniques*, Fig. 12.7, reproduced by permision of Oxford University Press.

Figure 30.6A: Courtesy of J.C.H. Spence, from Spence, J.C.H. and Zuo, J.M. (1992) *Electron Microdiffraction*, Fig. A1.3, reproduced by permission of Plenum Press.

Figure 30.6B: Courtesy of O.L. Krivanek, from Krivanek, O.L. (1988) in Buseck, P.R., *et al.* (Eds.) *High-Resolution Electron Microscopy and Associated Techniques*, Fig. 12.8, reproduced by permission of Oxford University Press.

Figure 30.7: Courtesy of S. McKernan.

Figure 30.8: Courtesy of Z.C. Lin, from Lin, Z.C. (1993) Ph.D. dissertation, Fig. 4.15, University of Minnesota.

Figure 30.9: Courtesy of O. Saxton, from Kirkland, A.I. (1992) in Hawkes, P.W. (Ed.) Signal and Image Processing in Microscopy and Microanalysis, *Scanning Microscopy* Supplement **6**, 139, Figs. 1, 2, 3, reproduced by permission of Scanning Microscopy International.

Figure 30.10: From Zou, X.D. and Hovmöller, S. (1993) *Ultramicroscopy* **49**, 147, Fig. 1, reproduced by permission of Elsevier Science B.V.

Figure 30.11A: From Kirkland, A.I., *et al.* (1995) *Ultramicroscopy* **57**, 355, Fig. 1, reproduced by permission of Elsevier Science B.V.

Figure 30.11B: From Kirkland, A.I., *et al.* (1995) *Ultramicroscopy* **57**, 355, Fig. 3, reproduced by permission of Elsevier Science B.V.

Figure 30.12A–C: From Kirkland, A.I., *et al.* (1995) *Ultramicroscopy* **57**, 355, Fig. 8, reproduced by permission of Elsevier Science B.V.

Figure 30.13: Courtesy of O.L. Krivanek, from Krivanek, O.L. and Fan, G.Y. (1992) in Hawkes, P.W. (Ed.) Signal and Image Processing in Microscopy and Microanalysis, *Scanning Microscopy* Supplement **6**, p. 105, Fig. 4, reproduced by permission of Scanning Microscopy International.

Figure 30.14: Courtesy of O.L. Krivanek, from Krivanek, O.L.and Fan, G.Y. (1992) in Hawkes, P.W. (Ed.) Signal and Image Processing in Microscopy and Microanalysis, *Scanning Microscopy* Supplement **6**, p. 105, Fig. 5, reproduced by permission of Scanning Microscopy International.

Figure 30.15: Courtesy of U. Dahmen, from Paciornik, S., *et al.* (1996) *Ultramicroscopy* **62**, 15, Fig. 1, reproduced by permission of Elsevier Science B.V.

Figure 30.16: Courtesy of U. Dahmen, from Paciornik, S., *et al.* (1996) *Ultramicroscopy* **62**, 15, Fig. 5, reproduced by permission of Elsevier Science B.V.

Figure 30.17: Courtesy of A. Ourmazd, from Kisielowski, C., *et al.* (1995) *Ultramicroscopy* **58**, 131, Figs. 2–4, reproduced by permission of Elsevier Science B.V.

Figure 30.18: Courtesy of A. Ourmazd, from Kisielowski, C., *et al.* (1995) *Ultramicroscopy* **58**, 131, Figs. 8, 10, 12, reproduced by permission of Elsevier Science B.V.

Figure 30.19A–D: Courtesy of A. Ourmazd, from Ourmazd, A., *et al.* (1990) *Ultramicroscopy* **34**, 237, Figs. 1, 1, 2, 5, reproduced by permission of Elsevier Science B.V.

Figure 30.20A–F: Courtesy of A. Ourmazd, from Kisielowski, C., *et al.* (1995) *Ultramicroscopy* **58**, 131, Figs. 15a–f, reproduced by permission of Elsevier Science B.V.

Figure 30.21: Courtesy of U. Dahmen, from Paciornik, S., *et al.* (1996) *Ultramicroscopy*, in press, Fig. 2, reproduced by permission of Elsevier Science B.V.

Figure 30.22: From King, W.E. and Campbell, G.H. (1994) *Ultramicroscopy* **56**, 46, Fig. 1, reproduced by permission of Elsevier Science B.V.

Figure 30.23: From King, W.E. and Campbell, G.H. (1994) *Ultramicroscopy* **56**, 46, Fig. 6, reproduced by permission of Elsevier Science B.V.

Figure 30.24: Courtesy of M. Rühle, from Möbus, G., *et al.* (1993) *Ultramicroscopy* **49**, 46, Fig. 6, reproduced by permission of Elsevier Science B.V.

Figure 30.25: From Thon, F. (1970) in Valdrè, U. (Ed.) *Electron Microscopy in Materials Science*, p. 571, Fig. 36, reproduced by permission of Academic Press.

Figure 30.26: Courtesy of J. Heffelfinger.

Chapter 31

Figure 31.2: Courtesy of R. Sinclair, from Sinclair, R., *et al.* (1981) *Met. Trans.* **12A**, 1503, Figs. 13, 14, reproduced by permission of ASM International.

Figure 31.4A,B: From Marcinkowksi, M.J. and Poliak, R.M. (1963) *Phil. Mag.* **8**, 1023, Figs. 15a, b, reproduced by permission of Taylor and Francis.

Figure 31.4C,D: Courtesy of J. Silcox, from Silcox, J. (1963) *Phil. Mag.* **8**, 7, Fig. 7, reproduced by permission of Taylor and Francis.

Figure 31.5: Courtesy of A.J. Craven, from Buggy, T.W., *et al.* (1981) *Analytical Electron Microscopy—1981*, p. 231, Fig. 5, reproduced by permission of San Francisco Press.

Figure 31.6D,E: Courtesy of NSA Hitachi Scientific Instruments Ltd. and S. McKernan.

Figure 31.7: Courtesy of R. Sinclair.

Figure 31.8: From Kuesters, K.-H., *et al.* (1985) *J. Cryst. Growth* **71**, 514, Fig. 4, reproduced by permission of Elsevier Science, B.V.

Figure 31.9: Courtesy of of M. Mallamaci.

Figure 31.10A: From De Cooman, B.C., *et al.* (1985) *J. Electron Microsc. Tech.* **2**, 533, Fig. 1, reproduced by permission of John Wiley & Sons Inc.

Figure 31.10B: Courtesy of S.M. Zemyan.

Figure 31.10C–E: Courtesy of B.C. De Cooman.

Figure 31.12: Courtesy of G. Thomas, from Bell, W.L. and Thomas, G. (1972) in Thomas G., *et al.* (Eds.) *Electron Microscopy and Structure of Materials*, p. 53, Fig. 28, reproduced by permission of University of California Press.

Figure 31.13: Courtesy of K.-R. Peters, from Peters, K.-R. (1984) in Kyser, D.F., *et al.* (Eds.) *Electron Beam Interactions with Solids for Microscopy, Microanalysis and Lithography*, p. 363, Fig. 1, reproduced by permission of Scanning Microscopy International.

Figure 31.14: Courtesy of R. McConville, from Williams, D.B. (1987) *Practical Analytical Electron Microscopy in Materials Science*, 2nd Edition, Fig. 3.11, reproduced by permission of Philips Electron Optics.

Figure 31.15: Courtesy of Philips Electronic Instruments, from Williams, D.B. (1987) *Practical Analytical Electron Microscopy in Materials Science*, 2nd Edition, Fig.

3.10, reproduced by permission of Philips Electron Optics.

Figure 31.16: Courtesy of H. Lichte, from Lichte, H. (1992) *Scanning Microscopy*, p. 433, Fig. 1, reproduced by permission of Scanning Microscopy International.

Figure 31.17: Modified from Lichte, H. (1992) *Ultramicroscopy* **47**, 223, Fig. 1.

Figure 31.18: Modified from Tonomura, A.: Courtesy of NSA Hitachi Scientific Instruments Ltd.

Figure 31.19A–C: From Tonomura, A. (1992) *Adv. Phys.* **41**, 59, Fig. 29, reproduced by permission of Taylor and Francis.

Figure 31.19D: From Tonomura, A. (1987) *Rev. Mod. Phys.* **59**, 639, Fig. 41, reproduced by permission of The American Physical Society.

Figure 31.20A: From Tonomura, A. (1992) *Adv. Phys.* **41**, 59, Fig. 38, reproduced by permission of Taylor and Francis.

Figure 31.20B: From Tonomura, A. (1992) *Adv. Phys.* **41**, 59, Fig. 42, reproduced by permission of Taylor and Francis.

Figure 31.20C: From Tonomura, A. (1992) *Adv. Phys.* **41**, 59, Fig. 44, reproduced by permission of Taylor and Francis.

Figure 31.21: Courtesy of R. Sinclair, from Sinclair, R., *et al.* (1994) *Ultramicroscopy* **56**, 225, Fig. 5, reproduced by permission of Elsevier Science B.V.

Chapter 32

Figure 32.1: Courtesy of J.E. Yehoda, from Messier, R and Yehoda, J.E. (1985) *J. Appl. Phys.* **58**, 3739, Fig. 1, reproduced by permission of the American Institute of Physics.

Figure 32.2: Courtesy of S.M. Zemyan.

Figure 32.3B: Courtesy of JEOL USA Inc.

Figure 32.4: Modified from Woldseth, R. (1973) *X-ray Energy Spectrometry*, Kevex Instruments.

Figures 32.5–8: Courtesy of S.M. Zemyan.

Figure 32.10: Courtesy of S. M. Zemyan, from Zemyan, S. and Williams, D.B. (1995) in Williams, D.B., *et al.* (Eds.) *X-ray Spectrometry in Electron Beam Instruments*, Fig. 12.9, reproduced by permission of Plenum Press.

Figure 32.11: Courtesy of S. M. Zemyan, from Zemyan, S. and Williams, D.B (1995) in Williams, D.B., *et al.* (Eds.) *X-ray Spectrometry in Electron Beam Instruments*, Fig. 12.10, reproduced by permission of Plenum Press.

Figure 32.12A: Courtesy of J.J. Friel, from Mott, R.B. and Friel, J.J. (1995) in Williams, D.B., *et al.* (Eds.) *X-ray Spectrometry in Electron Beam Instruments*, Fig. 9.8, reproduced by permission of Plenum Press.

Figure 32.12B: Courtesy of C.E. Lyman, from Lyman C.E., *et al.* (1994) *J. Microsc.* **176**, 85, Fig. 9, reproduced by permission of the Royal Microscopical Society.

Figure 32.13: Courtesy of S.M. Zemyan.

Figure 32.14: Courtesy of S.M. Zemyan.

Figure 32.15: Courtesy of D.E. Newbury, from Newbury, D.E. (1995) in Williams, D.B., *et al.* (Eds.) *X-ray Spectrometry in Electron Beam Instruments*, Fig. 11.18, reproduced by permission of Plenum Press.

Figure 32.16: Courtesy of J.I. Goldstein, from Goldstein, J.I., *et al.* (1992) *Scanning Electron Microscopy and X-ray Microanalysis*, 2nd edition, Fig. 5.3, reproduced by permission of Plenum Press.

Figure 32.17A,B: Courtesy of S.M. Zemyan.

Chapter 33

Figure 33.1: From Williams, D.B. (1987) *Practical Analytical Electron Microscopy in Materials Science*, 2nd Edition, Fig. 4.5a, reproduced by permission of Philips Electron Optics.

Figure 33.2: Courtesy of W.A.P. Nicholson, from Nicholson, W.A.P., *et al.* (1982), *J. Microsc.* **125**, 25, Fig. 4, reproduced by permission of the Royal Microscopical Society.

Figure 33.3: From Williams, D.B. (1987) *Practical Analytical Electron Microscopy in Materials Science*, 2nd Edition, Fig. 4.30, reproduced by permission of Philips Electron Optics.

Figure 33.4: Courtesy of S.M. Zemyan.

Figure 33.5A,B: Courtesy of G. Cliff, from Cliff, G. and Kenway, P.B. (1982) *Microbeam Analysis—1982*, p. 107, Figs. 5, 4, reproduced by permission of San Francisco Press.

Figure 33.6: Modified from Williams D.B. and Goldstein, J.I. (1981) in Heinrich, K.F.J., *et al.* (Eds.) *Energy-Dispersive X-ray Spectrometry*, p. 346, Fig. 7a, NBS.

Figure 33.7: Courtesy of S.M. Zemyan.

Figure 33.8: Courtesy of S.M. Zemyan.

Figure 33.9A: Courtesy of S.M. Zemyan.

Figure 33.9B: Courtesy of K.S. Vecchio, from Vecchio, K.S. and Williams, D.B. (1987) *J. Microsc.* **147**, 15, Fig. 1, reproduced by permission of the Royal Microscopical Society.

Figure 33.10A: Courtesy of S.M. Zemyan.

Figure 33.10B: Courtesy of S.M. Zemyan, from Zemyan, S. and Williams, D.B. (1994) *J. Microsc.* **174**, 1, Fig. 6, reproduced by permission of the Royal Microscopical Society.

Chapter 34

Figures 34.1–5: Courtesy of S.M. Zemyan.

Chapter 35

Figures 35.1–4: Courtesy of S.M. Zemyan.

Figure 35.5A: From Williams, D.B. (1987) *Practical Analytical Electron Microscopy in Materials Science*, 2nd Edition, Fig. 4.20, reproduced by permission of Philips Electron Optics.

Figure 35.5B,C: Courtesy of S.M. Zemyan.

Figure 35.6: Courtesy of S.M. Zemyan.

Figure 35.7: From Wood, J.E., *et al.* (1984) *J. Microsc.* **133**, 255, Figs. 2, 8, reproduced by permission of the Royal Microscopical Society.

Figure 35.8: From Bender, B.A., *et al.* (1980) *J. Amer. Ceram. Soc.* **63**, 149, Fig. 1, reproduced by permission of the American Ceramic Society.

Figure 35.10: Courtesy of S. Vivekenand and K. Barmak.

Figure 35.11A: Courtesy of J.A. Eades, from Christenson, K.K. and Eades, J.A. (1986) *Proc. 44th EMSA Meeting*, p. 622, Fig. 2, reproduced by permission of MSA.

Figure 35.12: Courtesy of R. Ayer, from M. Raghavan, *et al.* (1984) *Met. Trans.* **15A**, 783, Figs. 4, 11, reproduced by permission of ASM International.

Figure 35.13: Courtesy of A.W. Nicholls, from Nicholls, A.W. and Jones, I.P. (1983) *J. Chem. Phys.* **44**, 671, Figs. 3, 6a, reproduced by permission of The American Physical Society.

Figure 35.14A: Courtesy of V.J. Keast.

Figure 35.14B: Courtesy of J.R. Michael, from Michael J.R. and Williams, D.B. (1987) *Met. Trans.* **15A**, Fig. 7, Reproduced by permission of ASM International.

Figure 35.15: Courtesy of C.E. Lyman, from Lyman C.E. (1986) *Ultramicroscopy* **20**, 119, Figs. 1b, c, reproduced by permission of Elsevier Science B.V.

Tables 35.1, 35.2: From Williams, D.B. (1987) *Practical Analytical Electron Microscopy in Materials Science*, 2nd Edition, Tables 4.2a, b, reproduced by permission of Philips Electron Optics.

Tables 35.3A,B: From Wood, J.E., *et al.* (1984) *J. Microsc.* **133**, 255, Tables 9,11, reproduced by permission of the Royal Microscopical Society.

Chapter 36

Figure 36.1A,B: Courtesy of V.J. Keast.

Figure 36.2: Courtesy of J.R. Michael, from Williams D.B., *et al.* (1992) *Ultramicroscopy* **47**, 121, Fig. 1, reproduced by permission of Elsevier Science B.V.

Figure 36.3: Courtesy of J.R. Michael, from Williams D.B., *et al.* (1992) *Ultramicroscopy* **47**, 121, Fig. 2, reproduced by permission of Elsevier Science B.V.

Figure 36.4: Courtesy of R. Ayer, from Michael, J.R., *et al.* (1989) *J. Microsc.* **160**, 41, Fig. 2, reproduced by permission of the Royal Microscopical Society.

Figure 36.5.: Courtesy of R. Ayer, from Michael, J.R., *et al.* (1989) *J. Microsc.* **160**, 41, Figs. 3, 4, reproduced by permission of the Royal Microscopical Society.

Figure 35.8A,B: From Williams, D.B. (1987) *Practical Analytical Electron Microscopy in Materials Science*, 2nd Edition, Fig. 4.27, reproduced by permission of Philips Electron Optics.

Figure 36.9: Courtesy of C.E. Lyman, modified from Lyman C.E. (1987) in Kirschner, J., *et al.* (Eds.) *Physical Aspects of Microscopic Characterization of Materials*, p. 123, Fig. 1, Scanning Microscopy International.

Figure 36.10: Courtesy of C.E. Lyman, modified from Lyman C.E. (1987) in Kirschner, J., *et al.* (Eds.) *Physical Aspects of Microscopic Characterization of Materials* p. 123, Fig. 7, Scanning Microscopy International.

Chapter 37

Figure 37.1: Courtesy of R.F. Egerton, modified from Egerton, R.F. (1996) *Electron Energy-Loss Spectroscopy in the Electron Microscope*, 2nd edition, Fig. 2.2, Plenum Press.

Figure 37.2: Courtesy of Gatan Inc.

Figure 37.3: Courtesy of D.C. Joy, from Joy, D.C., *et al.* (Eds.) (1986) *Principles of Analytical Electron Microscopy*, Fig. 5, p. 259, reproduced by permission of Plenum Press.

Figure 37.4: Courtesy of J. Bruley.

Figure 37.5: Courtesy of Gatan Inc.

Figure 37.6: Courtesy of J.A. Hunt, from Hunt, J.A. and Williams, D.B. (1994) *Acta Microsc.* **3**, 1, Fig. 7, reproduced by permission of the Venezuelan Society for Electron Microscopy.

Figure 37.10: Courtesy of R.F. Egerton, from Egerton, R.F., *et al.* (1993) *Ultramicroscopy* **48**, 239, Fig. 2, reproduced by permission of Elsevier Science B.V.

Figure 37.11: Courtesy of O.L. Krivanek, modified from Krivanek, O.L, *et al.* (1991) *Microsc. Microanal. Microstruct.* **2**, p. 315, Fig. 8.

Chapter 38

Figures 38.1–3: Courtesy of J. Bruley.

Figure 38.4: Courtesy of O.L. Krivanek, modified from Ahn, C.C. and Krivanek, O.L (1983) *EELS Atlas* p. iv, Gatan Inc.

Figure 38.5: Courtesy of J. Bruley, modified from Joy, D.C. (1986) in Joy D.C., *et al.* (Eds.) *Principles of Analytical Electron Microscopy*, p. 249, Fig. 8, Plenum Press.

Figure 38.6: Courtesy of J. Bruley.

Figure 38.7: Courtesy of C.E. Lyman, from Lyman C.E. (1987) in Kirschner, J., *et al.* (Eds.) *Physical Aspects of Microscopic Characterization of Materials*, p. 123, Fig.

2, reproduced by permission of Scanning Microscopy International.

Figure 38.8: Courtesy of D.C. Joy, modified from Joy, D.C. in Hren, J.J., *et al.* (Eds.) (1979) *Introduction to Analytical Electron Microscopy*, p. 235, Fig. 7.6, Plenum Press.

Figure 38.9: Courtesy of M. Kundmann.

Figure 38.10: Courtesy of J.A. Hunt, from Hunt, J.A. and Williams, D.B. (1994) *Acta Microsc.* **3**, 1, Fig. 5, reproduced by permission of the Venezuelan Society for Electron Microscopy.

Figure 38.11: Courtesy of J.A. Hunt, from Hunt, J.A. and Williams, D.B. (1994) *Acta Microsc.* **3**, 1, Fig. 4, reproduced by permission of the Venezuelan Society for Electron Microscopy.

Figure 38.12: Courtesy of J. Bruley, from Hunt, J.A. and Williams, D.B. (1994) *Acta Microsc.* **3**, 1, Fig. 6, reproduced by permission of the Venezuelan Society for Electron Microscopy.

Table 38.2: Courtesy of R.F. Egerton, from Egerton R.F. (1996) *Electron Energy-Loss Spectroscopy in the Electron Microscope*, 2nd edition, p. 157, Table 3.2, reproduced by permission of Plenum Press.

Chapter 39

Figure 39.1: Courtesy of J. Bruley.

Figure 39.2: Courtesy of K. Sato and Y. Ishiguro, modified from Sato K. and Ishiguro Y. (1996) *Materials Transactions Japan Institute of Metals* **37**, 643, Figs. 1 and 7.

Figure 39.3: Courtesy of J. Bruley.

Figure 39.4: Courtesy of J.A. Hunt, from Hunt, J.A. and Williams, D.B. (1994) *Acta Microsc.* **3**, 1, Fig. 14, reproduced by permission of the Venezuelan Society for Electron Microscopy.

Figure 39.5: Courtesy of J.A. Hunt, from Williams, D.B. and Goldstein, J.I. (1992) *Microbeam Anlaysis* **1**, 29, Fig. 11, reproduced by permission of VCH.

Figure 39.6: Courtesy of J.A. Hunt, from Hunt, J.A. and Williams, D.B. (1994) *Acta Microsc.* **3**, 1, Fig. 17a, reproduced by permission of the Venezuelan Society for Electron Microscopy.

Figure 39.7: Courtesy of J.A. Hunt, from Hunt, J.A. and Williams, D.B. (1994) *Acta Microsc.* **3**, 1, Fig. 17b, reproduced by permission of the Venezuelan Society for Electron Microscopy.

Figure 39.8: Courtesy of D.C. Joy, from Joy, D.C. in Joy, D.C., *et al.* (Eds.) (1986) *Principles of Analytical Electron Microscopy*, p. 288, Figs. 6,7, reproduced by permission of Plenum Press.

Figure 39.9: Courtesy of R.F. Egerton, modified from Egerton R.F. (1993) *Ultramicroscopy* **50**, Fig. 6.

Figure 39.10: From Liu D.R. and Williams, D.B. (1989) *Proc.*

Roy. Soc. (London) **A425**, Fig. 7, reproduced by permission of The Royal Society.

Figure 39.11: Courtesy of J. Bruley.

Figure 39.12: Courtesy of D.C. Joy, from Joy, D.C. in Joy, D.C., *et al.* (Eds.) (1986) *Principles of Analytical Electron Microscopy*, p. 293, Fig. 10, reproduced by permission of Plenum Press.

Figures 39.13, 14, 15: Courtesy of J. Bruley.

Figure 39.16: Courtesy of J.A. Hunt, from Hunt, J.A. and Williams, D.B. (1994) *Acta Microsc.* **3**, 1, Fig. 16, reproduced by permission of the Venezuelan Society for Electron Microscopy.

Figure 39.17: Courtesy of R.F. Egerton, from Egerton, R.F. (1996) *Electron Energy-Loss Spectroscopy in the Electron Microscope*, 2nd edition, Fig. 1.11, reproduced by permission of Plenum Press.

Figure 39.18: Courtesy of O.L. Krivanek, modified from Krivanek, O.L., *et al.* (1991), *Microsc. Microanal. Microstruc.* **2**, 257, Fig. 5.

Chapter 40

Figure 40.4: Courtesy of N.J. Zaluzec, modified from Zaluzec, N.J. (1982) *Ultramicroscopy* **9**, 319, Fig. 3.

Figure 40.5A,B: Courtesy of J. Bruley.

Figure 40.5C: Courtesy of P.E. Batson, from Batson, P.E. (1993) *Nature* **366**, 727, Fig. 1, reproduced by permission of Macmillan Journals Ltd.

Figure 40.7: Courtesy of R. Brydson, modified from Hansen, P.L., *et al.* (1994) *Microsc. Microanal. Microstruc.* **5**, 173, Figs. 1, 2.

Figure 40.8: Courtesy of J. Bruley.

Figure 40.9: From Williams D.B. and Edington, J.W. (1976) *Acta Met.* **24**, 323, Fig. 7, reproduced by permission of Elsevier Science B.V.

Figure 40.9C: Courtesy of A.J. Strutt.

Figure 40.10A,B: Courtesy of J. Bruley.

Figure 40.10C: Courtesy of J.A. Hunt from Hunt J.A. and Williams, D.B. (1991) *Ultramicroscopy* **38**, 47, Fig. 11, reproduced by permission of Elsevier Science B.V.

Figure 40.11: Courtesy of J. Bruley, from Müllejans H., *et al.* (1993) *Electron Microscopy and Analysis—1993*, p. 62, Fig. 4, reproduced by permission of the Institute of Physics Publishing.

Figures 40.12, 13: Courtesy of J.A. Hunt.

Figure 40.14: Courtesy of J. Bruley and J. Mayer.

Figure 40.15: Courtesy of J.C.H. Spence and J. Mayer, from Mayer, *et al.* (1991) *Proc. 49th EMSA Meeting*, p. 787, Figs. 2, 3, San Francisco Press, reproduced by permission of MSA.

Table 40.1: From Williams D.B. and Edington, J.W. (1976) *J. Microsc.* **108**, 113, Table 1, reproduced by permission of The Royal Microscopical Society.